MICHAEL J. BRADSHAW

THE GEOPOLITICS OF ENERGY SYSTEM TRANSFORMATION

Managing the Messy Mix

First published in Great Britain in 2026 by

Bristol University Press
University of Bristol
1–9 Old Park Hill
Bristol
BS2 8BB
UK
t: +44 (0)117 374 6645
e: bup-info@bristol.ac.uk

Details of international sales and distribution partners are available at bristoluniversitypress.co.uk

© Michael J. Bradshaw 2026

DOI: 10.51952/9781529247312

The digital PDF and ePub versions of this title are available open access and distributed under the terms of the Creative Commons Attribution-NonCommercial-NoDerivatives 4.0 International licence (https://creativecommons.org/licenses/by-nc-nd/4.0/) which permits reproduction and distribution for non-commercial use without further permission provided the original work is attributed.

British Library Cataloguing in Publication Data
A catalogue record for this book is available from the British Library

ISBN 978-1-5292-4729-9 paperback
ISBN 978-1-5292-4730-5 ePub
ISBN 978-1-5292-4731-2 ePdf

The right of Michael J. Bradshaw to be identified as author of this work has been asserted by him in accordance with the Copyright, Designs and Patents Act 1988.

All rights reserved: no part of this publication may be reproduced, stored in a retrieval system, or transmitted in any form or by any means, electronic, mechanical, photocopying, recording, or otherwise without the prior permission of Bristol University Press.

Every reasonable effort has been made to obtain permission to reproduce copyrighted material. If, however, anyone knows of an oversight, please contact the publisher.

The statements and opinions contained within this publication are solely those of the author and not of the University of Bristol or Bristol University Press. The University of Bristol and Bristol University Press disclaim responsibility for any injury to persons or property resulting from any material published in this publication.

Bristol University Press works to counter discrimination on grounds of gender, race, disability, age and sexuality.

Cover design: blu inc
Front cover image: iStock/mnphotography

For Sally and Lily

Contents

List of Figures, Tables and Boxes		vi
List of Acronyms		viii
About the Author		ix
Preface		x
one	Introduction	1
two	Energy Futures, Geopolitics, and the Messy Mix	19
three	Fossil Fuel Geopolitics	45
four	The Resource Curse, Rentier States, and Unburnable Carbon	69
five	The Geopolitics of the Low-Carbon Transition	95
six	Managing the Messy Mix	129
References		139
Index		175

List of Figures, Tables and Boxes

Figures
1.1	Global greenhouse gas emissions by sector, 2021	5
1.2	Global energy consumption by source, 1800–2023	7
2.1	The emissions gap and climate change risks	25
2.2	The balance between fossil and nonfossil energy consumption in the IEA's 2024 World Energy Outlook	30
2.3	The geopolitics of energy system transformation	41
2.4	The three horizons and energy system transformation	43
3.1	Crude oil prices ($US) – the influence of geopolitical and economic events	51
3.2	Europe's gas import infrastructure	61
3.3	The Dutch Title Transfer Facility (TTF) gas price	62
3.4	EU imports of Russian pipeline gas, Russian LNG, US LNG, and Total LNG	64
4.1	Countries where 'oil rent' is more than 10% of GDP, 2021	71
4.2	Oil production per capita, 2023	72
4.3	Major oil producing states, 2023	73
4.4	GCC states – hydrocarbons' share of government revenues, 2013 and 2023	86
5.1	The geographical concentration of selected critical materials in 2023	102
5.2	The geographical concentration of refined products, 2020 and 2024	104
5.3	Geographical concentration of clean technology manufacturing capacity, 2023	113

| 5.4 | A simplified EV battery supply chain | 115 |

Tables

1.1	Effects of climate change	6
1.2	Global primary energy consumption by source, 1900–2023	8
1.3	Key indicators by World Bank income groups, 2023	10
2.1	The IEA's GEC Model 2024 scenarios	29
2.2	Alternative energy futures	31
3.1	Fossil fuel and renewable energy geopolitical issues	48
4.1	Major traders of crude oil and oil products, 2023	74
4.2	Vulnerability of petrostates	75
4.3	Oil rents as a percentage of GDP, GCC States, 1990–2021	85
4.4	GCC, fiscal breakeven oil price	87
4.5	World oil production and price by IEA's 2024 WEO scenarios	88
5.1	Mapping minerals with relevant low-carbon technologies	99
5.2	NUPI/IRENA composite list of critical materials for the energy transition	100
5.3	Key electricity security terms and definitions	121
6.1	The critical geopolitical challenges of energy system transformation	130

Boxes

2.1	Summary of scenario narratives	32
4.1	Policy prescription to address the resource curse	79
4.2	The lexicon of unburnable carbon	90

List of Acronyms

APS	Announced Pledges Scenario
BECCS	Bioenergy with Carbon Capture and Storage
BRI	Belt and Road Initiative
BRICS	Brazil, Russia, India, China, South Africa
CCS	Carbon Capture and Storage
COP	Conference of the Parties
COVID	Coronavirus Disease
DAC	Direct Air Capture
EST	Energy System Transformation
EV	Electric Vehicle
FSRU	Floating Storage Regasification Unit
GCC	Gulf Cooperation Council
GDP	Gross Domestic Product
GECF	Gas Exporting Countries Forum
GHG	Greenhouse Gas
GPN	Global Production Network
GWP	Global Warming Potential
IEA	International Energy Agency
IMF	International Monetary Fund
IPCC	Intergovernmental Panel on Climate Change
IRENA	International Renewable Energy Agency
LNG	Liquefied Natural Gas
MENA	Middle East and North Africa
NDC	Nationally Determined Contribution
NZE	Net-Zero Emissions Scenario
OPEC	Organization of the Petroleum Exporting Countries
PV	Photovoltaic
SDG	Sustainable Development Goal
UNEP	United Nations Environment Programme
UNFCCC	United Nations Framework Convention on Climate Change
WEO	World Energy Outlook

About the Author

Michael J. Bradshaw is Professor of Global Energy in the Strategy and International Business Group at Warwick Business School (WBS), University of Warwick. Before joining WBS, he worked at the University of Leicester and before that at the University of Birmingham. He is trained as a Human Geographer and holds a PhD from the University of British Columbia (Canada). He has served as Vice-President of the Royal Geographical Society (with IBG). He is also a Fellow of the Academy of Social Sciences. Recently, he completed a two-year secondment with Shell Scenarios as Senior Scenario Planner (2022–24). He is presently Associate Fellow at the Environment and Society Centre, Chatham House. He authored *Global Energy Dilemmas* (2014), which examined the links between energy security, globalization, and climate change. He is a co-editor of *Global Energy: Issues, Potentials and Policy Implications* (2015), and a co-author of *Energy and Society* (2018) and *Natural Gas* (2020). He is now retired from full-time work at WBS but continues to teach on their MBA programme and research for the UK Energy Research Centre (UKERC). His ongoing work with the UKERC includes a new project on the geopolitics of fossil fuel phase-out.

Preface

This book reflects over 25 years of research into the geopolitics of energy. My aim is threefold: first, to introduce a framework – the Messy Mix – to help navigate the complex geopolitical challenges posed by the transformation of the global energy system. Second, to serve as a bridge connecting the work of international organizations, business, think tanks, the media, and academia. Third, to integrate a wide range of issues from diverse perspectives into a holistic analysis of the geopolitics of energy system transformation.

One of my reviewers suggested that I describe the book as a guide, rather than a study or analysis. A dictionary definition of a guide would be a book that gives you the most important information about a particular subject. That is precisely my intent. To aid in this process, this guide is supported by an extensive reference list, and where possible, I have chosen open-access sources. Thanks to funding from Warwick Business School, this guide is available as an open-access PDF and e-Pub. It is also complemented by a website accessible via my WBS webpage (www.wbs.ac.uk/about/person/michael-bradshaw/).

This book is aimed at a wide readership, including academics and students interested in the geopolitics of energy, as well as those in government involved with energy and climate policy, individuals in business seeking to develop strategies to navigate the Messy Mix, and anyone with an interest in the role of energy and geopolitics in addressing climate change.

My academic journey has been supported by a wide range of institutions, organizations, and many individuals. Within the limits of this brief preface, it is impossible to name them all. Regarding this particular project, I am particularly grateful to David Elmes at WBS, with whom I teach an MBA module on Managing Sustainable Energy Transitions. Much of my thinking for this guide has been tested on various MBA cohorts. I have also gained from working with several PhD students, especially Tahani Almabadi, whose work on rentier states has been invaluable. Thanks also to Caroline Kuzemko in the Department of Politics and International Studies at Warwick (PAIS). The University of Warwick funded the research behind this guide through a period of academic leave. I also acknowledge UKRI and the UK Energy Research Centre (UKERC), who have supported my work over the past 15 years and continue to do so. Special thanks to the current and former Directors at UKERC, Rob Gross and Jim Watson, for recognizing the importance of research on geopolitics and its impact on the UK's energy transition in a global context. Post-doctoral researchers have supported various UKERC projects, and I am grateful to all of them. Particular mention goes to Mathieu Blondeel, who has led many of our recent publications. My research at UKERC has benefited from close collaboration with Gavin Bridge in Geography at Durham, as well as the modelling expertise at the UCL Energy Institute, particularly with Steve Pye and James Price. I have also enjoyed many years working with the Oxford Institute for Energy Studies, and I thank Jonathan Stern, Jim Henderson, and Jack Sharples for their assistance over the years. Most recently, I had the opportunity to spend two years on a secondment with Shell Scenarios as a Senior Scenario Planner. Thanks to Shell's Chief Political Analyst, Geraldine Wessing, for making this possible. I hope I have done justice to the benefits of foresight thinking. I joined the scenarios team at a particularly challenging moment, Spring 2022, and I thank everyone I worked with. It was a memorable experience.

Finally, thanks to those who helped in the production of this guide: various anonymous reviewers whose comments have helped improve the final product; Kerry Allen for drawing all the excellent figures; Louis Fletcher, in PAIS, who proofread the final manuscript – any errors remain my responsibility – and Izzie Green and Ellen Pearce at Bristol University Press.

ONE

Introduction

> Transitioning away from fossil fuels in energy systems, in a just, orderly and equitable manner, accelerating action in this critical decade, so as to achieve net zero by 2050 in keeping with the science. (UNFCCC, 2023, 4)

At COP-28 in Dubai in November 2023, the world committed to transitioning away from fossil fuels. I began writing this book a year later, shortly after COP-29 in Azerbaijan – which did not ratify the previous COP's commitments – and following President Trump's victory in the US elections. Since taking office in January 2025, President Trump has announced the US's withdrawal from the Paris Agreement, reversed support for clean-energy technologies, adopted an energy policy of 'Drill, baby, drill,' and initiated a global trade war. Despite his promise to end Russia's war on Ukraine overnight, peace remains elusive. A growing transatlantic divide is evident, as the US questions its willingness to provide a security guarantee for Europe. In the Middle East, prospects for a lasting peace between Israel and Hamas in Gaza appear limited. In Asia, China's President Xi Jinping has been flexing China's military might and consolidating his position as the leader of an alternative world order. According to the Institute for Economics and Peace's latest Global Peace Index, 56 conflicts are ongoing – the highest number since World War II (MOD, 2025). The UK's *Strategic Defence Review 2025* notes that 'The world is more volatile and more uncertain than at any time in the past 30 years, and it is changing rapidly' (MOD, 2025, 26). However, it is measured, we are in an era of heightened geopolitical risk. Simultaneously, the physical impacts of climate change are increasingly visible. Still, the world remains far from a path that would limit global warming this century to less than 2 °C, and as close as possible to 1.5 °C – the target set in the 2015 Paris Agreement. It is also true that the less action we take, the greater the likelihood of increased conflict due to the physical effects of climate change, and the higher the eventual cost of climate action (Stern, 2014). Despite all this, in some parts of the world, a populist 'greenlash' is currently underway, lambasting both the costs of the low-carbon energy transition and the necessity of net-zero policies. In other regions, concerns about energy security are accelerating efforts to transform the energy system to reduce reliance on fossil fuels.

This book examines the relationship between geopolitics and the pace of the transformation of the global energy system, shifting from the current dominance of fossil fuels to one reliant on low-carbon energy sources and carriers. The purpose of this first, introductory chapter is threefold: first, to assess the role of the energy system in the climate challenge; second, to explain the nature of energy system transformation (EST); and third, to explore the debate around the pace of transformation.

1.1 Climate change and the energy system

This guide begins with the premise that the evidence for climate change is unequivocal and that the primary cause of global warming is human activity, primarily the increased emission of greenhouse gases (GHGs) into the atmosphere. When I teach on this issue, I draw upon two main sources: the reports of the Intergovernmental Panel on Climate Change (IPCC, 2025a, np) and the climate change website of the US National Aeronautics and Space Administration (NASA, 2025, np). The IPCC is the United Nations body responsible for assessing the science related to climate change. Since its creation in 1988, it has conducted a series of 'Assessment Rounds' (AR) of 'the state of knowledge of the science of climate change; the social and economic impact of climate change, and potential response strategies and elements for inclusion in a possible future international convention on climate' (IPCC, 2025b, np). The most recent of these – AR 6 – was completed in 2023, with the publication of the *Summary for Policymakers*. Planning is now underway for the next round. The headline conclusion of AR 6, reached with a high degree of confidence, is that

> Human activities, principally through emissions of greenhouse gases, have unequivocally caused global warming, with global surface temperature reaching 1.1°C above 1850–1900 in 2011–2020. Global greenhouse gas emissions have continued to increase, with unequal historical and ongoing contributions arising from unsustainable energy use, land use and land-use change, lifestyles and patterns of consumption and production across regions, between and within countries, and among individuals. (IPCC, 2023, 4)

The IPCC reports are not an easy read, and even the *Summary for Policymakers* is challenging to digest. However, it effectively conveys the current state of our understanding of the climate change challenge. The NASA climate change website is much easier to navigate if it remains online in the face of the Trump Administration's cuts to climate science. It is regularly updated and will answer all the questions you might have about the evidence for climate change, its causes, and consequences. Here, I am most interested in the role that the global energy system plays in climate change.

There are six GHGs, but the most significant are carbon dioxide (CO_2), methane (CH_4), nitrous oxide (N_2O), and the so-called F-gases. They all exist in the atmosphere for varying durations and have different global warming potentials (GWPs). Carbon dioxide is assigned a GWP of 1, and the potency of all other GHGs is measured relative to this baseline. It is a long-lived gas that remains in the atmosphere for hundreds of years. As a result, most of the CO_2 accumulated in the atmosphere today was emitted over an extended period. Methane, by comparison, remains in the atmosphere for about 12 years. But because it is a more potent absorber of heat than CO_2, its 20-year GWP is 84–87 times greater. Even over a 100-year horizon, despite its short lifetime, methane has a GWP 28–34 times that of CO_2 (Ekins, 2023, 27–28). According to Richie, Rosado, and Rosser (2024, np), in *Our World in Data*, a source I will reference throughout this guide, in 2023, CO_2 accounted for 75% of GHG emissions related to energy, and CH_4 for a further 20%. It is common to combine various GHG emissions into a single measure: carbon dioxide equivalent (CO_2e). There are two main ways in which the energy system emits GHGs into the atmosphere: first, when fossil fuels are combusted; and second, through leakage of GHGs – primarily methane – during the production, processing, and transport of fossil fuels until their point of combustion or processing. Given its high GWP, reducing the so-called 'fugitive emissions' of methane can have a significant short-term impact on global warming.

As Figure 1.1 demonstrates, the energy system is by far the largest source of anthropogenic GHG emissions. The figure offers just a snapshot of one year, but the dominance of the energy sector has been consistent since the 1800s and the Industrial Revolution (more on this later). Scientific evidence shows a clear causal link between the release of GHGs into the atmosphere and the rate of global warming. The US National Oceanic and Atmospheric Administration (NOAA, 2025) at the Mauna Loa Observatory in Hawaii has been measuring the concentration of CO_2 in the atmosphere since March 1958, when it was 315.7 parts per million (ppm). By July 2025, the level had reached 427.84 ppm. This, after 2024, marked the most significant annual increase since records began. Longer-term data from ice cores indicate that human activities have increased CO_2 concentrations by 50%

INTRODUCTION

Figure 1.1: Global greenhouse gas emissions by sector, 2021

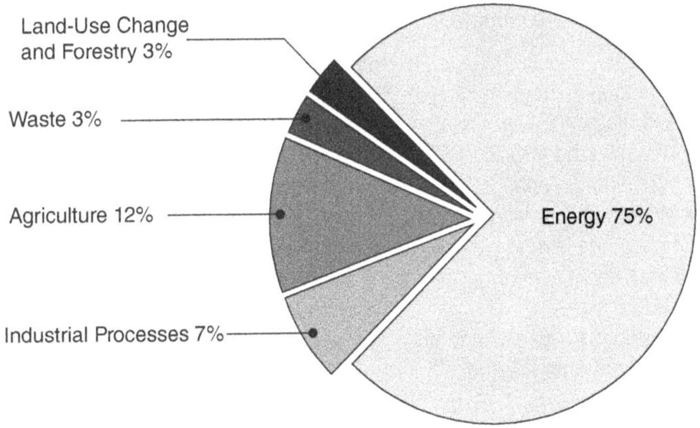

- Land-Use Change and Forestry 3%
- Waste 3%
- Agriculture 12%
- Industrial Processes 7%
- Energy 75%

Source: Calculated from ClimateWatch (2025, np)

since the Industrial Revolution, with a notable acceleration in recent decades.

Table 1.1, based on information from the UK Met Office, provides a summary of ongoing changes to the climate system and their effects. The physical risks of climate change are significant and becoming more obvious. The potential strategic impact of climate change is documented in various reports by defence and intelligence agencies (for example, NATO, 2024; MOD, 2024), where climate change complicates operations and could itself cause future conflicts. As a result, business as usual is no longer feasible as climate change speeds up. This should create a sense of urgency, emphasizing the need for swift action.

1.2 Energy system transformation

To appreciate the scale of the challenge we face, it is necessary to understand energy's past to contemplate its future. The Industrial Revolution marks a critical inflexion point in the relationship between human society and the energy system

Table 1.1: Effects of climate change

Changes to the climate system	Some of the effects of climate change
• Changes in the hydrological cycle • Warmer land and air • Warming oceans • Melting sea ice and glaciers • Rising sea levels • Ocean acidification • Global greening • Changes in ocean currents • More extreme weather	• Risk to water supplies • Conflict and climate migrants • Localized flooding • Flooding of coastal regions • Damage to marine ecosystems • Fisheries failing • Loss of biodiversity • Change in seasonality • Heat stress • Habitable region of pests expands • Forest mortality and increased risk of fires • Damage to infrastructure • Food insecurity

Source: Met Office (2025, np)

(Wrigley, 2010). Figure 1.2 illustrates the history of global energy consumption since 1800, with an unusual feature being the inclusion of an estimate for traditional biomass. Two things jump out from the data: the exponential growth of energy consumption since the turn of the 20th century, and the fact that the energy system's history has been characterized by the successive layering of new energy sources on top of one another, in a collectively growing stack (Smil, 2010).

By tabulating the underlying data from Figure 1.2, as in Table 1.2, it is possible to identify when a particular energy source became dominant. In 1900, coal accounted for 47.2% of global primary consumption, and oil a mere 1.5%; traditional biomass accounted for the remaining 50.4%. Despite the invention of the internal combustion engine in the late 1880s and its rapid adoption to power mobility from the early 1900s, coal remained king until well after World War II, when, from an aggregate global perspective, the age of oil finally arrived. In

Figure 1.2: Global energy consumption by source, 1800–2023

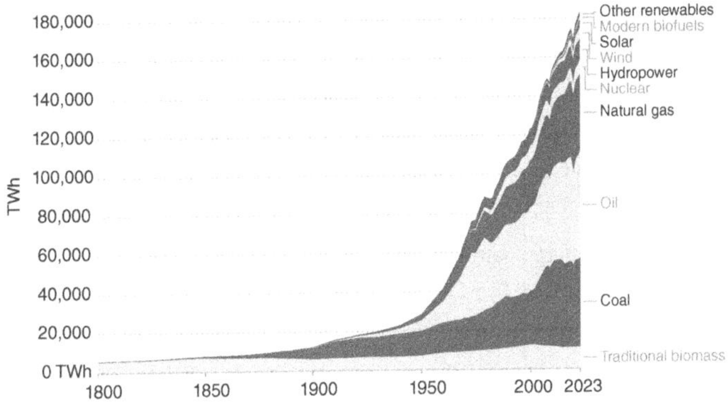

Legend (top to bottom): Other renewables, Modern biofuels, Solar, Wind, Hydropower, Nuclear, Natural gas, Oil, Coal, Traditional biomass

Source: Ritchie and Rosado (2024, np)

1950, coal accounted for 44% of global energy consumption, and oil 19%. It was only during the early 1960s that oil surpassed coal as the dominant energy source. By 1970, oil's share had grown to 40%, while coal had fallen back to 37%. However, percentages can be misleading; what matters for the climate is the absolute volume of GHGs being emitted through combustion and fugitive emissions.

Table 1.2 presents the absolute levels of consumption in terawatt-hours (TWh), which enables a comparison between different energy sources and carriers. This makes clear that while the 1970s saw a dramatic increase in the share of oil in the global energy mix, the physical amount of coal consumption continued to grow as well and reached a record level in 2024. Natural gas only gained traction outside the US in the late 1960s, spurred on by discoveries in the North Sea and West Siberia. Still, it was only in the 1990s that gas was permitted in power generation in the European Union (EU), prompting a 'dash for gas'. By the turn of the millennium, all three fossil fuels were well established, with oil dominant at 35%, followed by coal at 22.3%, and natural gas in third place at 19.4%. In total, these three fossil fuels accounted for nearly 77% of primary

Table 1.2: Global primary energy consumption by source, 1900–2023

(TWh)	1900	1950	1960	1970	1980	1990	2000	2010	2023
Traditional Biomass	6111	7500	8889	9444	10,000	11111	12500	11667	11111
Coal	5728	12603	15442	17087	20878	25916	27441	41988	45565
Oil	181	5444	11097	26673	35561	37608	42983	48,058	54564
Natural Gas	64	2092	4472	9615	14,237	19481	23994	31594	40102
Nuclear				224	2020	5677	7323	7373	6824
Hydropower	47	925	1914	3473	5120	6383	7826	9521	11014
Wind						11	93	961	6040
Solar						1	3	94	4264
Modern Biofuels				14	33	107	133	692	1318
Other Renewables				80	154	361	572	1,176	2428
Total	12131	28564	41814	66610	88003	106656	122868	153124	183230

Source: Ritchie and Rosado (2024, np)

energy consumption in 2000. This might seem lower than expected; this is because the estimate for traditional biomass, which is not included in the measure of commercial energy consumption, is factored in.

Figure 1.2 and Table 1.2 show that there has been a significant growth in the consumption of renewable energy (wind, solar, modern biofuels, and other renewables) in the last two decades. Taken together, they have grown from a mere 801 TWh in 2000 to 14,050 TWh in 2023, and their share has increased from less than 1% to nearly 8%. Viewed in isolation, this is an impressive growth rate. The problem is that fossil fuel consumption also increased by 45,813 TWh. Thus, there is no sign of a decline in the growth of fossil fuel consumption – quite the opposite. As with previous energy transitions, renewables are currently being added to, rather than displacing, growing fossil fuel demand (York and Bell, 2018; Fressoz, 2024; Yergin, Orszag, and Arya, 2025).

So far, the discussion has been at the global scale; clearly, there are significant regional and country-level differences in the history of energy, economic development, and GHG emissions. Table 1.3 presents some key indicators to inform the discussion. It uses the World Bank's country classification based on per capita income levels, which divides the world into four groups: high income (more than $14,005), upper-middle income ($4,516–14,005), lower-middle income ($1,146–$4,515), and low income (less than $1,145). The classification has evolved as countries have developed. In 1987, 30% of countries were classified as low income, and 25% were classified as high income; today, only 12% are classified as low income, and 40% are classified as high income (Metreau, Young, and Eapen, 2024, np).

Table 1.3 clearly illustrates stark inequalities. High-income countries, relative to their share of the world's population, have a disproportionate share of the world's wealth, energy consumption, and emissions, and are responsible for the majority of cumulative CO_2 emissions. To put it bluntly, they

Table 1.3: Key indicators by World Bank income groups, 2023
(% of World Total)

	Population[1]	National income (Atlas method)[1]	Primary energy consumption[1]	GHG emissions (excluding LULCF)[2]	Cumulative CO_2 emissions[2]*
High income	17.5	64.0	39.6	29.6	62.9
Upper-middle income	35.1	28.2	44.6	40.3	28.1
Lower-middle income	38.3	7.3	12.3	12.8	6.2
Low income	9.0	0.5	3.5	17.3	0.6

*2022 data

Sources:
[1] World Bank (2025, np)
[2] Our World in Data (2025, np)

INTRODUCTION

have combusted fossil fuels, benefited from the economic growth this made possible, and are primarily responsible for the GHG emissions that are the cause of climate change today. It is also the case that the world's poorest countries have contributed very little to the climate problem, but all the science shows that they are the most exposed to its physical risks and the least able to adapt to them. This creates a moral imperative for the world's wealthiest countries to reduce their emissions as quickly as possible and to provide financial assistance to the poorer nations to aid in their energy transitions (and adapt to climate change). For many, this means providing universal access to modern energy. This notion of 'differentiated responsibilities based on respective capabilities and social and economic responsibilities' is enshrined in the 1992 UN Framework Convention on Climate Change. It was also at the heart of the arguments over finance at COP-29 in Azerbaijan in November 2024, as well as at many other COPs before that. The rapid growth of China's economic, energy, and climate footprint is contributing to the increasing role of upper-middle-income economies. In the coming decades, the vast majority of energy demand growth will be concentrated in the middle-income group as living standards continue to improve. Thus, our climate and energy destiny lies in what the International Monetary Fund (IMF) calls the emerging and developing economies, a term I will favour in this guide.

Around 2010, the World Energy Council (WEC, 2024) coined the phrase 'Energy Trilemma' to refer to the complex trade-offs between energy security, energy equity, and environmental sustainability, and this has since gained widespread traction. The WEC produces an annual trilemma index that ranks countries across different dimensions; however, not all countries share the same priorities. Building on my earlier work and utilizing a combination of the World Bank and IMF's methods for categorizing the global economy, it is possible to identify three distinct energy transformation challenges (Bradshaw, 2014).

For developed high-energy societies (those with high incomes), the challenge is to rapidly decarbonize incumbent energy systems to achieve net-zero CO_2 emissions by 2050. Net-zero acknowledges that we need not reduce anthropogenic GHG emissions to absolute zero if we use 'negative emissions technologies' to draw down a volume of emissions from the atmosphere, equal to the volume of emissions we continue to produce. This includes nature-based solutions, such as tree planting (afforestation) and land-use change, as well as technical solutions, including carbon capture and storage (CCS) and the direct air capture of CO_2 (DAC). However, the likely contribution of CO_2 storage, for example, is uncertain, as there are both geological and technoeconomic limitations to its growth (Zhang, Jackson, and Krevor, 2025).

For emerging economies (upper/middle income), the challenge is to secure the energy needed to fuel economic growth and improve living standards, while limiting and then reducing GHG emissions (China has committed to reaching a peak in carbon emissions by 2030 and achieving net-zero by 2060, while India has pledged to reach net-zero by 2070). For developing economies (lower-middle income and low income), the challenge is to ensure sustainable energy access for all (World Bank Sustainable Development Goal 7), and a future based on sustainable prosperity while adapting to climate change. The key point here is that different parts of the world have different priorities regarding economic development, energy security, and climate change, as well as varying capabilities to address these challenges. Therefore, there is no single understanding of the problem, and no 'one size fits all solution'. Likewise, the energy history of the developed world should not be seen as a blueprint for the rest of the world.

Before leaving this discussion, it is necessary to clarify the distinction between transformation and transition (Kuzemko et al, 2024). Energy system transformation (EST) involves shifting the energy system away from reliance on fossil fuels to one mainly powered by low-carbon sources like renewable energy and electrification. At the heart of this process are two distinct transitions. First, a 'high-carbon transition' that involves the phasing out of fossil fuels as the dominant form of primary energy supply; and second, a low-carbon transition that consists of building out a low-carbon energy system. As already noted, previous transitions have layered new energy sources and services on top of the existing system; for the first time, we need to remove the incumbent system and replace it with a new one, and we need to do this in a very short period to avoid what climate scientists call 'catastrophic climate change'. The two-transition approach is essential, as there is a tendency to focus solely on the low-carbon transition, paying insufficient attention to the consequences of declining fossil fuel demand.

1.3 The speed of energy transitions

Before examining the range of possible energy futures presented in various scenarios and forecasts, I want to focus on the issue of the pace of energy transitions. This is a complex matter, on which a significant body of literature exists; here, I outline the contours of the debate, rather than focusing on the details. A good starting point is a White Paper by the World Economic Forum (WEF, 2019), which suggested that there are two distinct views on the speed of the energy transition, referred to as the gradual and rapid narratives. According to the WEF (2019, 6), the gradual narrative extrapolates: 'from the current patterns of policy, industry, consumption and investment'. The result is that 'the energy world of tomorrow will look roughly the same as that today … implying that the global energy system has an inertia incompatible with the Paris Agreement'. By comparison, the rapid narrative is that 'new energy technologies are rapidly supplying all the growth of energy demand, leading to peak fossil fuel demand in the course of the 2020s'; this is because 'current technologies and new policies will reshape markets, business models and patterns of consumption'.

The gradual narrative is supported by the work of energy historians, who maintain that energy transitions are complex and necessarily take a considerable amount of time. Grübler (2004) provided a long-term perspective on transitions in energy use and the relationship between population and economic growth; issues that are critical to our understanding of how things might unfold in the coming decades. Grübler (2012, 10–14) provided some crucial insights into the history of energy transitions. First, he emphasized the importance of energy end-use (demand) in driving these transitions. Second, he noted that rates of change tend to be slow, but not always. Finally, he argued that distinct patterns exist in the successful scaling up of technology systems (the so-called S-curve is discussed below). He also provided some cautionary tales, warning against moving 'too fast, too big and too early' (Grübler, 2012, 14), a view that retains contemporary resonance. In 2016, *Energy Research and Social Science* (Issue 22) hosted an open-access debate on energy transitions that involved researchers from both sides of the argument.

In that debate, Grübler, Wilson, and Nenmet (2016, 19) distinguished between an energy transition that they defined as a 'change in the state of an energy system' and what they called a 'grand energy transition' that brings about pervasive changes in the energy system: 'that affect multiple energy resources, carriers, sectors, and end use applications, often associated with the diffusion of "general purpose" technologies (e.g. steam engines, electricity)'. This issue is vital as those advocating a rapid transition tend to base their arguments on the experience of an individual country or a specific technology. Vaclav Smil (2016, 196) also participated in the debate and was very dismissive of the evidence in favour of more rapid change. He concluded that replacing the current global energy system based on fossil fuels is 'a task that will necessarily occupy us for generations to come'. Smil is a prolific and influential writer on energy transitions (Smil, 2010, 2017, 2024), and his message is consistent. He defines an energy transition as 'change in the composition (structure) of primary energy supply [emphasis in the original], the gradual shift from a specific pattern of energy provision to a new state of an energy system' (Smil, 2010, vii). There is much to be learnt from Smil's writings, in his 2010 volume *Energy Transitions: History, Requirements, Prospects* he concluded

> how far we will advance into the post-fossil future in three or four decades will not be determined only by the commitment to innovation but also our willingness to moderate our energy expectations and to have our energy uses following a more sensible direction, one that would combine reduced demand with a difficult, but eventually rewarding, quest for a civilization powered by renewable energy flows. (Smil, 2010, 153)

This echoes the importance of end-uses raised above and specifically highlights the importance of both demand reduction – using less energy – and improvements in energy efficiency – using less energy to perform a specific task. In many instances, the latter is a natural consequence of electrification, but demand reduction needs to be extended to the broader energy system. Curbing the growth of energy demand will reduce the scale of the low-carbon system that needs to be built. That said, Terzi and Fouquet (2023, 17) warn that what they call 'energy sobriety' and energy-efficiency improvements may have a limited impact due to the 'rebound effect' (also known as the 'Jevons paradox') and the growing energy demand resulting from economic growth.

In his latest offering, Smil's unequivocal conclusion is that 'after a quarter of a century of targeted energy transition, there has been no absolute global decarbonisation of energy supply' (Smil, 2024, 15). He evaluates the challenges, costs, and constraints involved in an accelerated transition and concludes that 'the world free of fossil carbon by 2050 is highly unlikely' (Smil, 2024, 33). The message from the gradualist school is clear: we face a grand energy transition on the scale of the Industrial Revolution, and the historical record suggests that it cannot be achieved in a matter of a few decades (Fouquet, 2024).

Those who dispute the gradual narrative do not dismiss the historical record. Instead, they suggest that the current energy transformation may be different and that more rapid change is both possible and necessary. Sovacool (2016), in the special journal issue mentioned above, suggested that numerous examples of rapid technological change and adoption exist, both in terms of specific technologies and particular country cases (see also Gros et al, 2018). At the heart of this debate lies the process of technological change and the S-curve, the lifecycle that successful innovations tend to follow, progressing from experimentation to scaling up, and then to standardization and widespread adoption. At the centre of the S-curve is a period of exponential growth, as costs fall rapidly and the new technology is adopted system-wide, before reaching saturation and plateauing, and eventually being replaced by the next innovation. Sovacool (2016, 206) acknowledged that the historical record suggests that when it comes to prime movers, innovations that change entire energy systems typically take a long time to implement; however, when it comes to end-users, changes can occur rapidly. Sovacool (2016, 210–11) maintained that three significant drivers make an accelerated transition possible. The first is scarcity; the current fossil fuel system is unsustainable because it is based on nonrenewable, stock resources that will eventually run out (more on this later). The second is climate change, and the need for a swift transition to mitigate the worst impacts of climate change. The third driver is that 'technological learning and innovation can result in new technologies and systems with the potential for exponential growth' (Sovacool, 2016, 210). Here, the cost of new technologies is fundamental, as is an understanding of the energy transition as a 'socio-technical process', and an entire subdiscipline has emerged around understanding transitions in 'socio-technical systems' (Geels and Turnheim, 2022). This is beyond the scope of this discussion. However, a combination of a deep understanding of historical energy transitions and an appreciation of the complexity of reconfiguring the current energy system provides fertile ground for developing approaches to accelerate EST (Sovacool et al, 2025).

Critical low-carbon technologies have been progressing along the S-curve, playing a vital role in reducing costs and accelerating deployment. The rate of progress is so significant that modellers at institutions like the IEA are regularly revising their forecasts upwards – for instance, solar power generation has consistently outperformed predictions. There is an increasing consensus that low-carbon power generation is gaining unstoppable momentum, as it remains the most cost-effective way to produce electricity in most contexts. Kingsmill Bond, a prominent 'S-curve optimist', along with his colleagues, contrasts two visions of the energy future: an 'old commodities narrative of business-as-usual' that predicts slow, costly, and painful reductions in fossil fuel demand; versus a 'new technology narrative of exponential and beneficial change' based on a shift to cheaper, faster, and more distributed systems (Bond, Butler-Sloss, and Walter, 2024, np). They classify these groups as the 'fossil gradualists' and the 'net-zero puritans'. They argue that 'the energy system is being transformed by the exponential forces of renewables, electrification, and efficiency' and that the 'orthodox view of slow change is wrong'. While they make a persuasive case, I believe the reality is more complex. Decarbonizing power generation is merely one part of a broader transformation that must also address intermittency and the need for long-duration energy storage, for example. Simultaneously, new infrastructure (the grid) must be developed to transmit low-carbon energy and facilitate the electrification of sectors such as transportation. This embodies the grand transition that the gradualist school discusses, which aligns with the concept of EST as used here.

1.4 Conclusions

So, how do we reconcile these competing narratives? There is no doubt that the historical record suggests that energy transitions, especially grand transitions, are long and drawn-out processes that take many decades. However, there is an argument that this transition is different; it is purposeful because it is driven by the need to address climate change. In 2024, $2 trillion was invested in low-carbon energy, compared to $1 trillion in fossil fuels (IEA, 2024a, 8). At COP-28, the world committed to tripling renewable energy capacity by 2030. Nonetheless, the task of phasing out fossil fuels should not be underestimated, given that they are crucial to every aspect of human society in the 21st century and that no energy transition has ever involved the subtraction of fossil fuels (though this is now happening locally in some parts of the world). The two narratives can be seen as different scenarios (the nature and role of scenarios are discussed in detail in the next chapter). The gradual transition results from exploratory thinking, where we create a future based on path dependency and the historical record, which suggests that inertia and strong vested interests will slow progress towards a low-carbon future. Conversely, the rapid narrative stems from normative thinking; it sets a target of a net-zero energy system by 2050 and works backwards, establishing the conditions needed to reach that goal without being constrained by historical precedent or geopolitical risks. Some may call this wishful thinking, but it provides a blueprint for accelerating EST. The next chapter explores energy scenarios more thoroughly before examining the role geopolitics plays in shaping the pace of this transformation.

TWO

Energy Futures, Geopolitics, and the Messy Mix

The previous chapter established that decarbonizing the energy system is vital for mitigating climate change and highlighted the significant uncertainty surrounding the pace of energy system transformation (EST). In this context, this chapter undertakes several tasks. Firstly, it explains the nature of scenario planning, a popular method for exploring the future of energy. Secondly, it examines the relationship between climate change and financial risk, which is essential for understanding how climate action affects the energy system. Thirdly, it presents the IEA's highly influential energy scenarios, which are used throughout this guide. Fourthly, it compares various industry scenarios to evaluate the influence of geopolitics on the pace of EST. This section also introduces a set of scenarios that frame the analysis presented in this guide. Lastly, the section examines the meaning of energy geopolitics and introduces the three horizons approach along with the concept of the 'Messy Mix'.

2.1 Scenario planning and energy's futures

Given that the current EST is unprecedented, it is no surprise that scenario planning is an essential approach for considering energy futures. Here, I explore the why, what, and how of scenario planning before examining two types of scenarios: the IEA's energy scenarios and a variety of scenarios produced by the energy industry. The aim is to understand the narratives they embody rather than to conduct a comparative quantitative analysis of scenario outcomes (IEF, 2025). Along the way, I also consider the financial risks associated with climate change and how these relate to the pace of transformation.

Why are scenarios helpful? When facing a high level of uncertainty, scenario thinking can offer an alternative to traditional forecasts, projections, or outlooks. The latter extrapolate from the recent past into the future based on quantitative analysis. Such thinking plays an important part but is most effective in the short term. In recent decades, energy markets have been particularly volatile; however, the industry must make large-scale investments, in the billions of dollars, that may take years to build and pay off over decades. In such a context, short-term forecasting is of limited use when thinking about long-term strategy (Bentham, 2014).

Scenario planning has its origins in military and strategic thinking, developed during the Cold War in the 1950s at organizations like the RAND Corporation in the US. In the energy sector, Shell has been a leading advocate of scenario thinking. As Pierre Wack, one of the pioneers of Shell Scenarios, observed 'The future is no longer stable; it has become a moving target. No single 'right' projection can be deduced from past behaviour. The better approach, I believe, is to accept uncertainty, try to understand it, and make it part of our reasoning' (Wack, 1985, 73). Shell has employed scenario planning since 1971, and the organization regards it as a fundamental part of its corporate culture. As we will see, scenario thinking is now prevalent across the climate and energy sectors, offering a structured approach to navigating extreme uncertainty.

So, what are scenarios, and how do you create them? Here, I draw on various writings from business and management, as well as the world of business consulting, and my own experience teaching with scenarios and working as a Senior Scenario Planner at Shell Scenarios (2022–4). Scenario planning is both an art and a science, and there are many approaches to it, to the extent that it has become part of a broader discipline and industry of foresight and futures thinking. To paraphrase a widely used business and management textbook by Goodwin and Wright (2014, 388–9), scenarios are pen pictures (narratives or stories) of a range of plausible futures. They are 'what if' stories of how the future might unfold, but plausibility is crucial. The different scenarios are built around a set of key uncertainties and certainties – inevitable truths – that result from rigorous research into the current situation and the identification of the key drivers of future change. But, as Wilson (2000, 24) so elegantly put it: 'However good our futures research may be, we shall never be able to escape from the ultimate dilemma that all our knowledge is about the past, and all our decisions are about the future.' This is especially relevant in the context of the earlier discussion of the pace of energy transitions; just because they have taken decades in the past does not necessarily mean that this will be the case in the future. The challenge is identifying those critical uncertainties that determine the pace of change.

Garvin and Levesque (2005) provided a step-by-step guide to formulating scenarios, in a process similar to that used by Shell Scenarios. It starts with a 'key focal issue' together with a clear definition of the scope and timeframe. The focus here is on the energy system, and the timeframe extends to 2050, which is a timeframe that is meaningful in terms of business and investment cycles. The first step is the identification of the 'driving forces', 'the themes and trends that affect, influence and shape the focal issue in fundamental ways' (Garvin and Levesque, 2005, 3). Next are the 'critical uncertainties', those driving forces that are likely to define or significantly change how the future unfolds. If we identify, say, the pace of innovation or the level of international cooperation as critical uncertainties about the future, we can then use these factors to help differentiate our scenarios. Often, these driving forces are paired to produce a simple 2×2 matrix that creates four different 'quadrants of uncertainty', which can be explored in more detail. The final step is to create an internally coherent set of scenarios that are 'plausible, alternative hypotheses about how the world might unfold, specifically designed to highlight the risks and opportunities facing a particular organisation or sector' (Garvin and Levesque, 2005, 3).

There are no right or wrong scenarios. Equally, there are no hard and fast rules for scenario planning, and different organizations have different approaches. Roxburgh (2009) provided some applicable 'rules of thumb', some of which are discussed below. He suggested that you should continually develop at least four scenarios, because if you create three, people will always pick the middle one as being the most likely. However, if you look at the history of Shell Scenarios, for example, they started with as many as six or seven, which proved too many, and then for a while settled on just two, but then turned to three in their Energy Transformation Scenarios – Waves, Islands, and Sky 1.5 (Shell, 2021). They then returned to two in their Energy Security Scenarios – Archipelagos and Sky 2050 (Shell, 2023). But in their recent update to the Energy Security Scenarios, they have renamed Sky 2050 as Horizon and added a third scenario – Surge – to capture the potential impact of artificial intelligence (AI) on the global energy system (Shell, 2025).

Roxburgh (2009, 8) suggested that in developing scenarios, it would be typical to identify three to five critical uncertainties that could then generate any number of 2×2 matrices. These can be combined to underpin your scenario narratives. He suggested that there should always be a base case, which is the most likely outcome. I am not sure about this. In the current context, we do not know what the most likely outcome is, but we do know that 'business as usual' is implausible as significant change is already underway. However, we can identify the most desirable outcome (achieving the goals of the Paris Agreement). The point of scenario planning, though, is to present a range of plausible futures, desirable or otherwise. Inevitably, people will still pick the scenario they like the most, the one that confirms their own beliefs or is likely to result in a more positive outcome for their organization, whatever that might be. At the same time, scenario thinking can allow for contrary voices that challenge 'groupthink'. At the end of the day, from an organization's perspective, whether it is a government department, a corporation, or any other form of organization, scenario planning aims to make better decisions by considering a range of possible outcomes.

Finally, three observations on the nature of scenario planning that build on our earlier discussion. First, as noted in the previous chapter, it is now commonplace to distinguish between 'exploratory scenarios' and 'normative scenarios' (Skea et al, 2021, 2). This distinction is significant in the energy and climate space, and there is some confusion surrounding it. In Shell's Energy Security Scenarios 2025, both Archipelagos (a fragmented world driven by security concerns) and Surge (a world where AI accelerates the pace of economic growth) are exploratory scenarios, while Horizon is a normative scenario that aims to deliver a net-zero energy system by 2050 (to explore the details, visit the Shell Scenarios website).

Second, the distinction between forecasts and scenarios is often a false dichotomy. Some scenarios are sets of projections based on a central forecast that explore key sensitivities to produce different outcomes or pathways. Such scenarios tend to be more technocratic, lacking the qualitative storytelling characteristic of true scenario thinking. These differences will become clear as I explore a range of scenarios, outlooks, and pathways.

Third, when using energy modelling to support scenario thinking, it is essential to acknowledge the assumptions underlying the models and the quality of the data they utilize. Because the future is uncharted territory, the models create possible futures based on a particular technology mix and set of price assumptions, as well as judgments on the pace of innovation and deployment. This can result in what Clift and Kuzemko (2024, 756) call the 'past as future presumption'.

Most energy models also have an inbuilt supply-side bias, and many have a particular appetite for solutions that rely heavily on negative carbon emissions technologies, which are not yet available at scale, such as biomass and carbon capture and storage (BECCS), and, more recently, direct air capture. Furthermore, while energy models are well calibrated to the incumbent fossil fuel system, they struggle to represent critical aspects of a future low-carbon energy system. In summary, all models are simplifications of reality and should be treated accordingly. This suggests a possible tension between the qualitative storytelling of scenario planning, which considers all possible futures and the models that support them, and the limitations these models impose on climate mitigation possibilities (Clift and Kuzemko, 2024).

2.2 The emissions gap and the financial risks of climate change

This section examines the disparity between the current pace of climate action and the pace necessary to meet the objectives of the Paris Agreement. It then considers the consequences of climate change as viewed by financial institutions. The *Emissions Gap Report* is produced annually by the United Nations Environment Programme (UNEP). It assesses the gap between current and projected GHG emissions, and the levels needed to meet the temperature goals of the Paris Agreement (UNEP, 2024). Figure 2.1 presents the findings of the 2024 report in graphical form. The historical trend in GHG emissions is self-explanatory. The top line shows the trajectory based on current policies, indicating a business-as-usual outcome. The following two lines relate to the nationally determined contributions (NDCs) made by the parties to the Paris Agreement – 194 states and the EU – that account for 98% of global GHG emissions. Under the Paris Agreement, NDCs must be updated, and their ambition increased every five years, most recently at COP-30 in Brazil. The two lines

Figure 2.1: The emissions gap and climate change risks

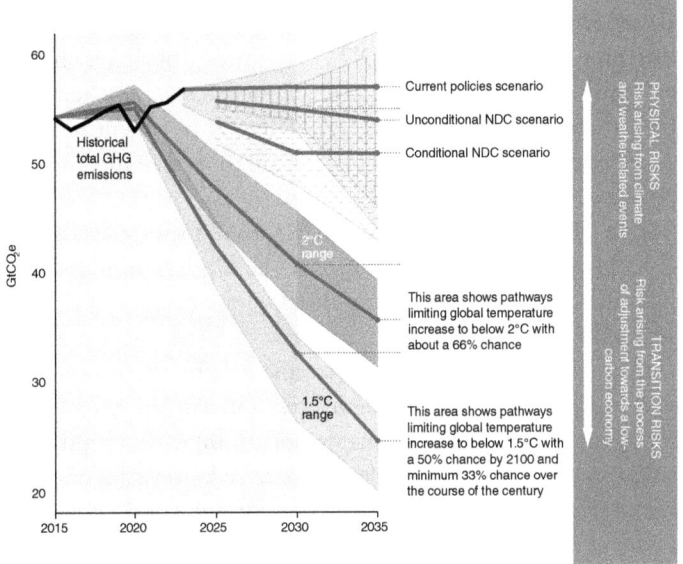

Source: UNEP, 2024, XVI

at the bottom represent the emissions pathways associated with 2 °C and 1.5 °C of warming, and the shaded areas depict the range of uncertainty. A significant gap exists between a future based on current policies and the 1.5 °C goal.

For clarification, an unconditional NDC is a climate action plan that a country can achieve without external financial assistance. In contrast, a conditional NDC can only be achieved with the help of external financial support. Developing countries can submit both forms of NDC, while developed countries can only submit unconditional NDCs. This links the level of climate action to the thorny issue of financial assistance for developing countries, which is a sticking point in the COP process, as the commitments of the developed countries have fallen short of expectations. Although the Paris Agreement has undoubtedly made a difference, the world is significantly off track to meet its goals. Just how far off track was made evident in 2022 at the launch of the IPCC Working Group III report on mitigation of climate change that noted 'limiting warming to around 1.5 °C (2.7 °F) requires global greenhouse gas emissions to peak before 2025 at the latest and be reduced by 43% by 2030' (IPCC, 2022a). There is an urgent need to accelerate the pace of progress if we are to reduce the physical risks associated with climate change.

The financial system identifies two types of climate risk: physical risk and transition risk. Understanding these risks is a crucial component of developing energy and climate scenarios (NGFS, 2024). Physical risks are the direct result of climate change – and include extreme weather events, such as storms, floods, and wildfires – that result in damage to property, infrastructure, and supply chains, for example. Such physical risks can be acute, for example, the increased severity of extreme weather events such as heat waves or chronic conditions, such as the gradual rise in sea level. These risks present considerable challenges to society, and they are often most acute in regions that are least able to adapt. In the context of energy security, these risks already affect the world's energy infrastructures, and their increasing frequency and impact are a cause for concern. For example, a recent study by China Water Risk concluded that 12 of the world's top 15 oil tanker terminals will be impacted by just 1 m of sea level rise (Tan et al, 2024). Physical risks also threaten low-carbon energy infrastructures, as hurricane-force winds can damage wind turbines and power grids. Meanwhile, periods of drought impact hydropower generation, and low river levels result in insufficient water to cool nuclear power stations (Mikellidou et al, 2018).

Transition risks are the result of climate action and include the consequences of carbon taxes, mandates for carbon disclosure, and the transition to renewable energy. A rapid transition could result in stranded assets whereby fossil fuel infrastructure and carbon-intensive industries are rendered unprofitable well before the end of their economic life (this is discussed in Chapter 4). In an extreme case, the asset could be stranded before it has recouped its investment, let alone provided a return. This is more of a concern for recent and new investments that have yet to reach cost recovery, rather than legacy assets that are fully paid off. However, there is also the possibility that low prices will not even cover production costs, as was the case with US LNG exports at the beginning of the COVID-19 pandemic in spring 2020. Figure 2.1 makes clear the relationship between the pace of the transition and the risk profile associated with climate change. A gradual transition results in more significant physical risks, with the more chronic risks becoming apparent after 2050.

In contrast, accelerated action will limit physical risks but increase the level of transition risks as the economy rapidly shifts away from fossil fuel and carbon-intensive activities. In this context, those with significant assets in the incumbent energy system favour a gradual transition, while those most impacted by physical climate risks demand a more rapid transition. This discussion highlights why the pace of EST is the most critical uncertainty.

2.3 The IEA's energy scenarios

The IEA was established in 1974 to assist OECD member states in coordinating their response to the oil crisis at that time, leading to the creation of the Strategic Petroleum Reserve (SPR). In recent years, the organization has expanded its scope beyond energy security to address the challenges of the energy transition. Since 2015, it has also increased its membership beyond the OECD. It is, without question, the most influential international body dealing with energy issues, and governments, businesses, civil society, and academic researchers widely use its scenarios. Since 1993, the IEA has provided a variety of medium- to long-term projections. From 2021 onwards, it has developed a hybrid modelling approach to offer a comprehensive analysis of how to transition the world's energy system to net-zero by 2050. These scenarios are a key part of its annual *World Energy Outlook* (IEA, 2024, a), which is now published ahead of the main yearly COP meeting. The scenarios have evolved, and since 2015, they have integrated the objectives of the Paris Agreement with the UN's Sustainable Development Goal 7: Affordable and Clean Energy. In 2021, the IEA (2021b) published *Net Zero by 2050: A Roadmap for the Global Energy Sector*, outlining a pathway to achieve net-zero by 2050 and limit global temperature rise to 1.5 °C. It was controversial because it stated 'Beyond projects already committed as of 2021, there are no new oil and gas fields approved for development in our pathway, and no new coal mines or mine extensions are required' (IEA, 2021b, 21). Subsequently, it has refined its stance on new investment in oil and gas, stating that 'no new long lead-time conventional oil and gas projects are required, and no new coal mines or coal mine lifetime extensions are needed either' (IEA 2024b, 239) in the net-zero pathway, which is now their normative scenario. Climate campaigners have used this to support the view that fossil fuels must stay in the ground; more on this later. Table 2.1 summarizes the three scenarios as presented in the 2024 World Energy Outlook (WEO).

The difference between STEPS and APS reveals an 'implementation gap' as APS assumes that all global climate commitments, net-zero targets and SDG-7 are realized on schedule. Thus, it represents the best we can hope for, given our current situation. The NZE scenario outlines what is needed to achieve a net-zero energy system by 2050, and the difference between this and APS is described as an 'ambition gap'. The NZE scenario is a normative scenario that works backwards from a defined outcome. At the same time, STEPS and APS are exploratory scenarios that establish different sets of starting considerations and consider where they may lead (IEA 2024b, 79). It is noteworthy that the IEA is re-introducing the 'Current Policies' scenario in its 2025 Outlook to reflect the fact that countries are currently failing to deliver on even

Table 2.1: The IEA's GEC Model 2024 scenarios

Name	Description
Stated policies (STEPS)	This scenario represents current policy settings, analysing energy-related policies in place and under development as of August 2024, on a sector-by-sector and country-by-country basis. It also considers planned manufacturing capacities for clean-energy technologies.
Announced Pledges (APS)	This scenario assumes that all global climate commitments, including nationally determined contributions (NDCs), long-term net-zero targets, and goals for universal access to electricity and clean cooking, as of August 2024, will be fully achieved on schedule.
Net-Zero Emissions by 2050 (NZE)	This scenario outlines a pathway for the global energy sector to achieve net-zero CO_2 emissions by 2050, relying solely on energy sector actions. It includes achieving universal access to electricity and clean cooking by 2030 and incorporates updated data as of 2024.

Source: IEA (2024b, np)

their stated policies, let alone the future commitments they have made. This response also addresses the IEA's critics, who argue that all the scenarios are implausible because they are too far removed from reality and what they consider a plausible pace of change (Mills and Atkinson, 2025).

Figure 2.2 illustrates the outcomes of the three scenarios. It makes clear that STEPS does not result in a system transformation by 2050, which would mean 2.4 °C of warming. In APS, the point at which nonfossil energy sources become dominant is not until the 2040s, whereas in NZE, this happens in the mid-2030s. The APS scenario is within the IPCC's range for a 2 °C level of warming, while the NZE scenario lies within the 1.5 °C range.

While the IEA's scenarios are highly influential, and the WEO 2024, alongside other IEA reports, are an essential

Figure 2.2: The balance between fossil and nonfossil energy consumption in the IEA's 2024 World Energy Outlook

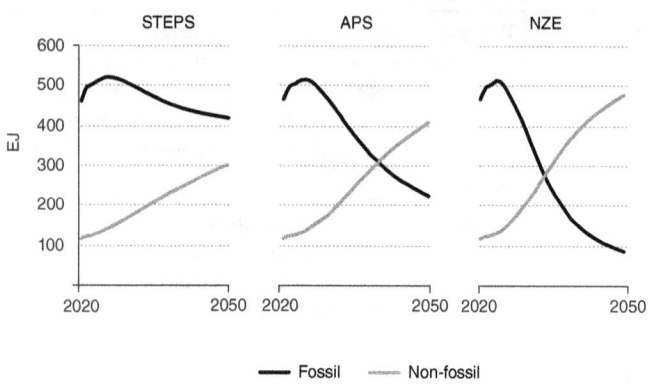

Source: IEA per. comm

source for this study, it is important to remember that they are scenarios generated by mathematical models that depend on their underlying assumptions and the data used.

2.4 Other futures are available

This section considers an assortment of energy scenarios developed by international oil companies and other organizations involved in the energy sector (Table 2.2). It does not claim to be representative, and I am most interested in the scenario range and nomenclature, as well as the essence of the different scenarios and their implications for climate change.

There are various approaches in terms of the number of scenarios and the range of temperature outcomes. Most have one or more gradual transition scenarios that are not Paris-aligned. All of these nonaligned scenarios are exploratory, extrapolating from the world as it exists today. Understandably, they all also have a normative Paris-aligned scenarios of less than 2 degrees, and some have more than one, including a net-zero 1.5 degrees scenario. When it comes to scenario names, there

Table 2.2: Alternative energy futures

Company	Name of reports and scenarios	Temperature outcome
Bloomberg NEF	**New Energy Outlook 2024** • Economic Transition • Net Zero	2.6 °C 1.75 °C
BP	**Energy Outlook 2024** • Current Trajectory • Net Zero	2.5 °C 1.5 °C
Energy Intelligence (EI)	**Energy Transition Macro Outlook 2025** • *Slow Down* • *Momentum* • *Velocity* • *Net Zero*	>2.5 °C ~2.5 °C <2.0 °C ~1.5 °C
Equinor	**Energy Perspectives 2024** • Walls • Bridges	>1.5 °C 1.5 °C
McKinsey & Co	**Global Energy Perspective 2024** • Slow Evolution • Continued Momentum • Sustainable Transformation • <1.5 °C	~2.6 °C ~2.2 °C ~1.8 °C ~1.5 °C
S & P Global	**The New Pragmatism** • Discord • Inflections • Green Rules	3.6 °C 2.5 °C 1.8 °C
Shell	**The 2025 Energy Security Scenarios** • Archipelagos • Surge • Horizon	2.2 °C ~2.0 °C <1.5 °C
TotalEnergies	**Energy Outlook 2024** • Trends • Momentum • Rupture	2.6–2.7 °C 2.2–2.3 °C 1.7–1.8 °C
Wood MacKenzie	**Energy Transition Outlook 2024** • Delayed Transition • Base Case • Country Pledges • Net Zero	3.0 °C 2.5 °C 2.0 °C 1.5 °C

are varying degrees of imagination, and three have a net-zero scenario. Box 2.1 presents an analysis of the texts that describe the scenarios with the highest degree of warming on the one hand, and the Paris-aligned scenarios on the other. ChatGPT 4.0 was tasked with analysing each text to produce a summary statement (Box 2.1).

Box 2.1: Summary of scenario narratives

A Gradual Transition: falling short. Bloomberg's (2024) **Economic Transition** scenario assumes no additional climate policies beyond current market trends. Emissions are projected to fall by 27% by 2050, as renewables surpass fossil fuels by 2030, driven by cost competitiveness and increased market access. BP's (2024) **Current Trajectory** reflects current global energy trends based on existing policies and pledges. Emissions peak in the mid-2020s, falling 25% below 2022 levels by 2050, but challenges persist in meeting climate targets. EI's (2025) **Slow Down** scenario depicts a weakening of policy support, stagnating clean tech deployment, and a return to fossil fuel investment due to geopolitical tensions, high costs, and weak economic conditions, resulting in slow emissions reductions Equinor's **Walls** scenario sees moderate progress influenced by economic growth and recent legislation such as the US Inflation Reduction Act and EU Green Deal. However, persistent geopolitical tensions and limited cooperation hinder deeper decarbonization. McKinsey's **Slow Evolution** scenario prioritizes energy affordability and security, leading to inconsistent regional decarbonization, reduced investment in clean energy, and severe environmental and social consequences. S & P's **Discord** scenario outlines a fractured, conflict-prone world where global cooperation erodes. Emissions are expected to remain dangerously high by 2050, despite limited progress in select countries. Shell's **Archipelagos** describes a world focused on national security, with an emissions peak in the 2020s. In the 2030s, decarbonization is sluggish, although energy security still promotes the uptake of some low-carbon technologies. TotalEnergies' **Trends** captures current policy and technology developments, especially in China, but infrastructure limitations and geopolitics stall further progress. Wood MacKenzie's **Delayed Transition** portrays a fragmented world where protectionism hinders the adoption of clean energy. Fossil fuels dominate, hydrogen stalls, and EV sales hit only 44% globally by 2050.

A Rapid Transition: a Paris-aligned future. Bloomberg's (2024) **Net Zero** emphasizes the urgent and large-scale effort required to cap warming at 1.75 °C and meet the Paris Agreement targets. It calls for a 93% reduction in power sector emissions by 2035, a 75% drop in oil demand by 2050, and a tripling of renewable energy capacity by 2030. All new car sales must be electric by 2034, with full electrification by 2046, while carbon capture and storage (CCS) and clean hydrogen scale dramatically. BP's (2024) **Net Zero** explores potential changes to the energy system under a 'what if' framework, targeting a 95% reduction in CO_2 emissions by 2050. EI's (2025) **Net Zero** assumes significant tightening of climate policies and societal shifts that support energy efficiency and low-carbon adoption, allowing a near −1.5 °C pathway with minimal overshoot. Equinor's (2024) **Bridges** scenario envisions a cooperative geopolitical environment, rapid decarbonization, and widespread deployment of renewables by 2030. McKinsey's (2024) **<1.5 °C scenario** emphasizes global cooperation, investment in emerging economies, and accelerated technology adoption to cut emissions by 50% by 2030 and reach net zero by 2050. S & P Global's (2024) **Green Rules** outlines an accelerated transition driven by climate and security concerns but falls short of achieving global net-zero by 2050. Shell's **Horizon** is a normative scenario targeting net-zero by 2050 and <1.5 °C by 2100, with public-led technology adoption and rapid solar expansion. TotalEnergies' **Rupture** scenario aims to limit warming to under 2 °C, requiring significant global decarbonization by 2040. Wood MacKenzie's **Net Zero** scenario assumes global cooperation and technological innovation to meet the 1.5 °C goal by 2100, with 91% EV uptake, 460 million tons of low-carbon hydrogen, and 6 billion tons of CCS deployed, despite a temporary overshoot in the early 2030s.

When it comes to scenarios that fall short of the Paris target, the common causes that delay technology deployment include a lack of policy ambition, prioritizing energy security over climate action, and fragmentation and geopolitical barriers. By contrast, the Paris-aligned scenarios rely on transformative action to ensure the rapid deployment of key technologies for managing emissions and decarbonizing electricity generation. In some scenarios, national security concerns prompt accelerated climate action to reduce reliance on fossil fuels. Of course, there is plenty of middle ground between these two extremes, and two scenarios cannot capture the complexity of the situation. What is clear is that geopolitics is implicated in the pace of change; a failure to resolve fragmentation and conflict frustrates cooperation and slows the pace of transformation. Therefore, a less fragmented and more cooperative world order is crucial for effective climate action. These scenarios predate the actions of the second Trump Administration. As many are annual exercises, it will be interesting to see how the baseline is reset in the wake of the US Government's actions to roll back climate action domestically, withdraw from the Paris Agreement, and disrupt global trade. All of this is both a symptom and a cause of a more fragmented world.

By combining the earlier discussion of the transition pace and the review of various scenarios, this study considers three possible scenarios. The first two, the rapid and the gradual, come from the WEF (2019) report on the pace of the energy transition and are also reflected in the discussion in Box 2.1. The third 'A Messy Transition' is a narrative space that sits between the two, forming a future that serves as the basis for this guide. That is not to say that it is the most likely; rather, it is an exploration of how geopolitics influences the pace of EST.

The three energy futures are as follows:

- *A Gradual Transition* in which the energy world of tomorrow will look roughly the same as that of today, implying that the global energy system has an inertia incompatible with the Paris Agreement (this is a world of significant and growing physical climate change risk).
- *A Rapid Transition* whereby current and new clean-energy technologies are rapidly supplying all the growth in energy demand and together with new policies will reshape markets, business models and patterns of consumption leading to a peak in fossil fuel demand during the 2020s putting the world on a net-zero path (this is a world of significant and growing transition risk).

- *A Messy Transition* that fails to manage the phase out of fossil fuel consumption alongside the build out of clean power generation, electrification, and improvements in efficiency, resulting in increased price volatility, a public backlash against the transition and a breakdown in international cooperation on climate change (this is a world of significant and persistent geopolitical risk).

2.5 The geopolitics of energy system transformation

This final section turns to geopolitics and defines the key terms that are used in this guide. It then links back to the earlier discussion of pace and the role of scenario thinking before introducing the three horizons approach and the 'Messy Mix'. Geopolitics and geopolitical risk play a significant role in any assessment of the current state of the world (World Economic Forum, 2025). However, the meaning of geopolitics is seldom clearly defined. We can think of geopolitics as an academic endeavour that sits between political geography and international relations, but it also describes a wide range of actions taken by states (Lehmann, 2017). The energy sector is also the focus of sanctions and embargoes – so-called statecraft – to influence the actions of states (Sovacool, Baum, and Low, 2023). From a business perspective, geopolitics refers to the actions of states that impact the functioning of markets, investments, and international trade. The term 'geoeconomics' is increasingly used to explain such developments, but this is an unnecessary complication, and I see such concerns as part of geopolitics. However, because the term 'geopolitics' has different meanings to different people, it is essential to clarify what I mean by 'geopolitics' in this context.

2.5.1 Defining geopolitics

Here, I begin by examining what political geographers have to say about the origins and nature of geopolitics. As a subfield of political geography, geopolitics originated at the end of the 19th century, an age of empire built on sea power. In the 20th century, it gained notoriety because of its links to Nazi expansionism. In the post-war period, it was deployed in the containment of the Soviet Union. Since the 1970s, energy and geopolitics, particularly concerning oil, have been closely associated, a relationship analysed in the next chapter. It is possible to find numerous definitions of geopolitics, but there is also an essential contemporary distinction between what many would call 'classical geopolitics' and 'critical geopolitics' (Kelly, 2016). Dittmer and Sharp (2014, 5) distinguished between the two approaches by stating that 'classical geopolitics saw geography as the "reality" that needed to be analysed in order to guide foreign policy, critical geopolitics saw language as the building blocks from which reality emerged'. Critical geopolitics is concerned with representations and imaginaries, such as President Trump's notion of 'Energy Dominance', rather than the logic of resource endowment and relative location. Taking Kelly's (2016) lead, this analysis leans towards a more classical approach but avoids the perils of geographical determinism by allowing for the power of discourse. Kelly (2016, 2) stated that 'Geopolitics rests upon the relative spatial positions of countries, regions, and resources as they may affect foreign policies and actions.' Dodds (2019, 3) has added that geopolitics has three qualities: 'First, it is concerned with questions of influence and power over space and territory. Second, it uses geographical frames to make sense of the world ... Third, geopolitics is future-oriented.' In the current context, this third quality is significant, as I am interested in how geopolitics influences and is in turn influenced by the pace and outcome of EST.

2.5.2 A geopolitics of global energy security

So much for geopolitics; what particular characteristics are associated with the notion of energy geopolitics? More often than not, energy geopolitics is closely tied to energy security. Again, academia has made the definition and measurement of energy security a complex matter (Azzuni and Breyer, 2018; Chester, 2010; Sovacool and Mukherjee, 2011; Strojny et al, 2023). Earlier, I alluded to the IEA's definition of energy security, which is 'the uninterrupted availability of energy sources at an affordable price'. There is a need to problematize energy security beyond a simple definition (Cherp and Jewell, 2014). However, for the moment, I will adhere to the IEA's definition, although I prefer energy services over energy sources, as we consume the services that energy provides. Energy transformation is about delivering the same services in different ways, and potentially with reduced energy consumption.

In 2008, the late Malcolm Wicks MP, a former Minister of State for Energy in the UK Government, produced a report on energy security at the request of then-Prime Minister Tony Blair (Wicks, 2009). In defining energy security, he believed that energy policy must aim at achieving three things: *physical security*, avoiding involuntary interruptions of supply; *price security*, providing energy at reasonable prices to consumers; and *geopolitical security*, ensuring the *state* [UK in the original] retains independence in its foreign policy through avoiding over-reliance on particular nations. This approach is particularly relevant in the current context, as it links energy security to the state's sovereignty and its ability to act independently. This has particular resonance in the context of events in Europe since Russia's full-scale invasion of Ukraine (Skalamera, 2023). Kivimaa (2024, 38) makes an essential distinction between 'internal energy security' that relates to the secure operation of the energy system, and 'external energy security' that relates to the broader security implications of the energy system. This reflects the fact that, recently, but not for the first time, there has been a 'securitization' of the energy system, extending beyond traditional concerns for availability and affordability. Much of the current commentary on the pace of climate action reflects a view that, due to the current energy crisis, energy security has become a more pressing concern than decarbonization. However, the ideal is to pursue policies that both improve energy security and reduce carbon emissions.

2.5.3 Rethinking energy geopolitics

Back in 2009, I defined the 'geopolitics of global energy security' as 'the influence of geographical factors, such as the distribution of centres of supply and demand, on state and non-state actions to ensure adequate, affordable, and reliable supplies of energy' (Bradshaw, 2009a, 1921). That definition reflected the concerns of the time, and as discussed in the next chapter, our understanding of geopolitics and energy security is still largely shaped by fossil fuels. In the academic literature, the challenge to rethink energy geopolitics and energy security has grown alongside the increasing importance of renewable energy, to such an extent that a separate body of work has emerged on the 'geopolitics of renewables' (Scholten and Bosman, 2016; Paltsev, 2016; Scholten, 2018). This work forms the basis of Chapter 5, which explores the geopolitics of the low-carbon transition. Regarding policy, the publication of a 2019 report by the International Renewable Energy Agency titled *A New World: The Geopolitics of the Energy Transformation* highlighted the necessity of considering the geopolitical implications of EST driven by renewables (IRENA, 2019). The introduction to that report stated that

> The accelerating deployment of renewables has set in motion a global energy transformation that will have profound geopolitical consequences. Just as fossil fuels have shaped the geopolitical map over the last two centuries, the energy transformation will alter the global distribution of power, relations between states, the risk of conflict, and the social, economic and environmental drivers of geopolitical instability. (IRENA, 2019, 12)

Subsequently, IRENA has published a series of reports on the geopolitics of the energy transition, which are referenced throughout this guide. Interestingly, by comparison, the IEA has avoided direct engagement with geopolitics. Given its origins, the IEA is much more comfortable framing issues in terms of risks to energy security. A 2022 IEA report for G20 on The *Security of Clean Energy Transitions* noted that

> energy security is evolving, and the extent and type of risks to energy supplies are broadening, requiring countries to anticipate and manage both existing and newly emerging energy security challenges. Accelerated transitions are likely to amplify both old and new security factors, requiring the bolstering of resilience and emergency response capacities to ensure the uninterrupted flow of affordable energy. (IEA, 2022, 25)

2.5.4 Geopolitics and energy system transformation

The academic work on the geopolitics of renewables tends to focus on how a future low-carbon energy system will differ from the incumbent fossil fuel system, suggesting that many of the geopolitical tensions and energy insecurities we experience today will be a thing of the past. Hopefully, that will turn out to be the case; however, the immediate challenge is how to transition from our current system, dominated by fossil fuels and its associated problems, to the low-carbon energy system of the future. Figure 2.3 visualizes the two transitions of EST and their associated geopolitical challenges. As explained later, the 'Messy Mix' refers to the fact that we are entering into a period where we face both the problems of the past and the challenges of the future.

The curves in the figure depict the IEA's 2024 Net Zero Scenario. As previously discussed, there is significant uncertainty regarding the pace of change; equally, there is concern that the transformation process should be 'just', an issue addressed in various contexts throughout this guide (Atkins, 2023). The figure also highlights the geopolitical challenges central to each transition (Vakulchuk, Overland, and Scholten, 2020; Blondeel et al, 2021). Building on the idea that EST will alter the nature of geopolitics, I define the geopolitics of EST as 'the global shifts in power dynamics, economic relationships, and strategic alliances driven by changes in how countries produce, distribute, and consume energy'. In a recent handbook on the geopolitics of the energy transition, the editor states that the focus of that volume 'is on how the energy transition affects energy geopolitics, not the opposite causality, i.e. how global political rivalry affects the speed and direction of the energy transition' (Scholten 2023, 8). By contrast, this study examines the complex interrelationship between the technical determinants of energy transitions and the influence of geopolitics on the pace of transformation (Blondeel et al, 2024; Kuzemko et al, 2024).

2.5.5 Three horizons and the 'Messy Mix'

Innovation is central to EST. The low-carbon transition mainly involves developing and deploying technologies that harness, store, manage, and transport low-carbon energy. Electrification is key to the transition, which shifts energy supply chains away from flows of fuels and towards the manufacture and deployment of technologies like solar panels and batteries. Nonetheless, hydrogen, biofuels, and synthetic fuels will be necessary in hard-to-abate sectors, such as aviation. As explained earlier, achieving net zero also depends on technologies that capture, utilize, and store carbon dioxide. While these can allow for the continued use of fossil fuels, this is only acceptable if broader environmental, social, and political issues related to fossil fuels are also addressed.

In this guide, I utilize the three horizons approach to system change (Sharpe et al, 2016; Sharpe, 2020). The three horizons approach is well established within futures thinking (Curry and Hodgson, 2008). Figure 2.4 illustrates the three horizons and applies the framework to this study of the geopolitics of the energy transition. It does not have a direct connection to the IEA's scenarios, although it depicts a transition pathway that would likely align with the Paris Agreement. Following Curry and Hodgson (2008, 2–3), Horizon 1 (H1) is 'the prevailing system as it continues into the future, which loses "fit" over time as its external environment changes'. Horizon 3 (H3) refers to 'ideas or arguments about the future of the system which are, at best, marginal in the present, but which may over time have the potential to displace the world of the first horizon, because they represent a more effective response to changes in the external environment'. Horizon 2 (H2) is 'an intermediate space in which the first and third horizons collide. This is the transitional space, which is typically unstable. It is characterised by clashes in values in which actors propose competing alternative paths to the future'.

Figure 2.3: The geopolitics of energy system transformation

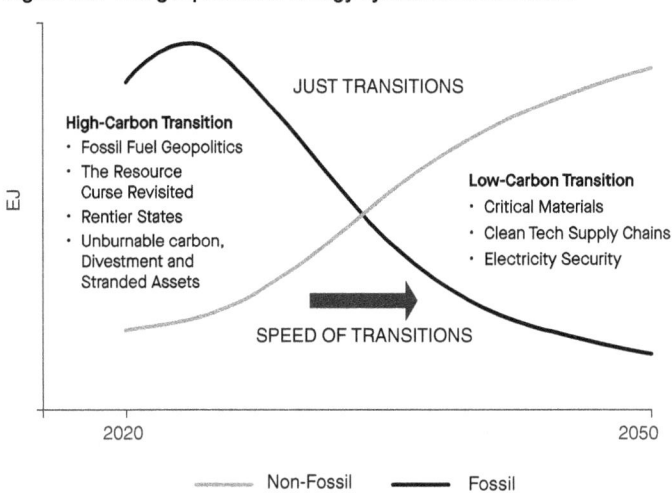

Source: Author

Curry and Hodgson's paper, published in 2008, was ahead of its time, as the authors usefully applied the framework to climate and energy security. For them, H1 signifies the current fossil fuel system, whose future faces threats from worries about climate change and resource availability; this is a world of decreasing abundance. In today's context, environmental sustainability seems to be a greater challenge than scarcity (Hone, 2023). H2 is 'the emerging short-term to medium-term future in which we know the limitations of our current position, but do not have the resources to respond effectively; there is little political agreement, the technology is immature, and so on' (Curry and Hodgson, 2008, 6). Finally, H3 'represents a future in which the limitations have been overcome', and it is a positive future based on renewable technologies and a mixture of macro (centralized) and micro (decentralized) systems. They also describe H2 as the 'Messy Mix', a period of 'transitional emergency' with increasing security breakdowns as some renewables mainstream and the endgame of oil and gas plays out. Much has changed since 2008; many renewable technologies are now mainstream, but the fossil fuel system remains stubbornly resistant to change. A recent IMF working paper (Espagne et al, 2023, 5) identifies what they call the 'unstable mid-transition period', when 'the fossil-based energy system will coexist with the emerging low-carbon energy system, while being increasingly impacted by increasing climate damages'. This perspective also aligns with the concept of 'energy pragmatism', which posits that the world will require a diverse range of energy sources and technologies, both low-carbon and higher-carbon options, for many decades to come (BlackRock, 2024).

I make no claims to the originality of the 'Messy Mix'. Figure 2.4 applies the framework developed by Curry and Hodgson (2008) to the current situation and the challenges of delivering a net-zero energy system. We are at the beginning of the end of the age of fossil fuels, and already in what I will call

Figure 2.4: The three horizons and energy system transformation

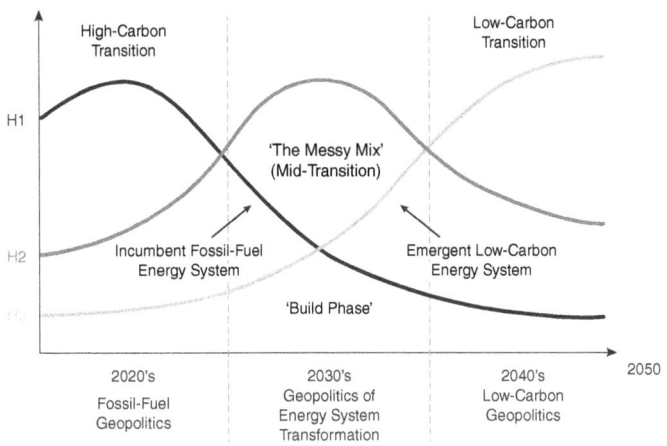

Source: Author

the 'build phase' of H3. The challenge is to secure the materials and to develop and install the technologies needed to create and operate the future low-carbon energy system. Eventually, the fossil fuel system will settle into a new steady state, and the low-carbon system will become the new incumbent. This is H3, when fossil fuel geopolitics will become a thing of the past. However, fossil fuels will continue to hold influence for decades to come. The literature on the geopolitics of the energy transition discusses 'winners and losers' (Van de Graaff and Sovacool, 2020, 70–2). The general direction of such analysis is that current fossil fuel producers will be the losers. At the same time, those well-positioned to benefit from the low-carbon transition will be the winners (Overland et al, 2019). In this context, this guide aims to understand the consequences of the 'Messy Mix' to enable EST in a more just, orderly, and equitable manner, thereby ensuring the development of the low-carbon energy system at the pace necessary to meet the goals of the Paris Agreement.

2.6 Conclusions

The coming years will increasingly resemble the 'Messy Mix', a world where we face all the geopolitical challenges of the current fossil fuel system and its externalities, further complicated by the effects of declining demand alongside the difficulties of the low-carbon transition's build phase. All this occurs against a backdrop of rising physical climate risks. The following two chapters analyse the geopolitics of fossil fuels (Chapter 3) and the consequences of the high-carbon transition for producer economies (Chapter 4). Failing to manage the transition risks linked to phasing out fossil fuels will lead to price volatility and geopolitical instability, conditions unfavourable to speeding up climate action. Chapter 5 explores the emerging geopolitical challenges associated with the build phase of the low-carbon transition. The final chapter summarizes the diverse geopolitical issues arising from the Messy Mix, which must be addressed to navigate an inherently disorderly transition.

THREE

Fossil Fuel Geopolitics

This is the first of two chapters that examine the geopolitics of fossil fuels. In 2024, oil, coal, and gas accounted for 86.6% of the total commercial energy supply (EI, 2025, 14). However, it is wise to consider the challenges that will arise as low-carbon alternatives replace fossil fuel demand. Put another way, to achieve an orderly transition away from fossil fuels, it is essential to manage the impact this will have on the stakeholders of the current fossil fuel energy system.

Their geography influences the geopolitics of fossil fuels, as they are located in specific countries, creating a degree of market concentration among so-called 'producer economies'. This results in dependence for importing nations and a market mechanism orchestrated by international trade that balances supply with demand. Nearly 40% of all shipping is dedicated to transporting oil, gas, and coal (McKibben, 2024). According to IRENA (2024, 29), 'The majority of countries are net fossil fuel importers, some 6.9 billion people – that is, 86% of the world's population – live in countries that are net fossil fuel importers.' Only 14% of the global population resides in countries that are net exporters of fossil fuels. The extent of international trade varies by fuel type, as does the level of market concentration. In 2023, the Asia-Pacific region accounted for 82% of global coal imports, with China as the largest importer and Indonesia the largest exporter (EI, 2024, 48–50). Högselius (2023, 71) observes that when it comes to coal, 'no single producing country or grouping of countries has ever been able to dictate market prices'. He suggests that this has fostered the perception that coal is somehow more secure than other fossil fuels. It appears that in times of crisis, many nations revert to coal, considering it more secure and affordable than oil and gas (Medzhidova, 2022). This partly explains why much of the literature on the geopolitics of fossil fuels concentrates on oil, with less emphasis on gas, and rarely on coal.

Things are much more complex when it comes to oil. The data suggest that in 2024, about 48% of crude oil production was traded internationally (EI, 2025, 22–34). According to the EIA (2023a), in 2022, approximately 96 countries produced oil, with five countries accounting for 52% of the global output. The 13 member states of OPEC (Organization of the Petroleum Exporting Countries) contributed 38% of global oil production, while the share of OPEC+, which includes Russia among others, was 59% of the total (EIA, 2023b). According to its Statute, OPEC's goal is

> to coordinate and unify the petroleum policies of its Member Countries and ensure the stabilization of oil markets in order to secure an efficient, economical, and regular supply of petroleum to consumers, a steady income to producers, and a fair return on capital for those investing in the petroleum industry. (OPEC, 2025, np)

Therefore, OPEC is at the forefront of managing the changing role of oil within the global energy system.

Natural gas differs again because it is more difficult to transport (Bradshaw and Boersma, 2020, 22–7). This explains why, in 2024, less than a quarter (21.5%) of global natural gas production was traded internationally, with 47.4% of that trade taking place via pipelines and 52.6% as seaborne liquefied natural gas (LNG) (EI, 2025, 37–42). Natural gas is converted into a liquid state by cooling it to extremely low temperatures, around −162 °C (or −260 °F). The members of the Gas Exporting Countries Forum (GECF) account for 69% of the world's gas reserves, 39% of marketed production, and

40% of global exports (GECF, 2025). However, they wield far less influence than OPEC, as key exporters Australia and the US are not members. Three countries – the US, Australia, and Qatar – represented 60% of global LNG trade in 2024 (EI, 2025, 43). The rise of LNG is a relatively recent development, and its geopolitical implications are discussed below in the context of the ongoing European gas crisis.

This chapter is divided into three sections. The first section explores the nature of fossil fuel geopolitics; the second examines the role of oil and the link between oil, conflict, and democracy; the third explains why the geopolitics of natural gas differs, using the current European gas crisis, prompted by Russia's war in Ukraine, as a case study. The chapter concludes by considering how the high-carbon transition will alter the nature of fossil fuel geopolitics.

3.1 The essentials of fossil fuel geopolitics

The literature on the geopolitics of energy system transformation often begins by comparing how the geopolitics of a future low-carbon system will differ from the current fossil fuel system. Table 3.1 is adapted from Vakulchuck, Overland, and Scholten's (2020) review of the literature on renewable energy and geopolitics, providing a foundation for this section and the following chapter. Their central premise is that fossil fuels cause more geopolitical issues than renewable energy.

The material characteristics of fossil fuels are crucial for understanding the geopolitical issues associated with the fossil energy system (Bridge and Le Billon, 2017, 6–39). The three fuels differ significantly, and none is homogeneous; there are varying grades and modes of extraction. However, all fossil fuels are nonrenewable stock resources, meaning their resource base is finite (on a human timescale), and they are depleted by consumption. This is unlike most minerals, which can be recycled, and unlike renewable sources such as hydro, wind, and solar power, which do not rely on fuels. Not long

Table 3.1: Fossil fuel and renewable energy geopolitical issues

Main issue	Fossil fuels	Renewable energy
Resource scarcity	Very significant	Only for critical materials
Importance of location	High	Moderate
Control over resources	Centralised	Decentralized
Geopolitical power	Asymmetric	Less asymmetric
International competition	High	Low
International interdependence	High	Low if renewables are domestic/high if imported
Security of supply	Highly important	Moderately important
Geopolitical tensions	Frequent	Opinions vary greatly
Conflict type	Large-scale and violent	Small-scale and nonviolent

Source: Selected from Vakulchuk, Overland, and Scholten (2020, 3)

ago, the concern was that, at current rates of consumption and production, the world would reach peak oil output (Aleklett and Campbell, 2003). Instead, the so-called 'Shale Revolution' in North America demonstrates how technological advancements can create new reserves by extracting oil and gas from impermeable shale rock through horizontal drilling and hydraulic fracturing (Evans, 2017). In less than 20 years, this technology has transformed the US from a major importer of oil and gas into the world's leading producer and exporter of LNG. This shift has had profound geopolitical consequences (O'Sullivan, 2017; Umbach, 2017; Yergin, 2020, 3–57). Among many factors, the availability of fossil fuels is affected mainly by price and geopolitics. Production costs for oil and gas vary greatly. When prices are low, investment slows, and production may decline as the most expensive wells are shut down. Conversely, when prices peak, there is often a surge in investment to boost output. One advantage of shale is its 'short cycle' nature, which allows it to be brought into production quickly and turned on or off according to market conditions (Bradshaw and Waterworth, 2018). This also makes shale oil particularly vulnerable to price fluctuations.

Global geopolitics has always played a significant role in the context of fossil fuels, but politics is becoming increasingly prominent, both in terms of international politics related to climate change and domestic politics in both producing and consuming states. Some reserve-holding states, such as Denmark, have decided to leave fossil fuels in the ground. At the same time, many fossil fuel-importing states have energy security and climate strategies that aim to reduce their consumption of fossil fuels. All of this suggests that physical scarcity is no longer the primary concern; instead, we should focus on who controls the lowest-cost fossil fuel reserves as demand decreases and prices decline. Who will extract the last barrel?

Fossil fuels, particularly oil, are seen as sources of geopolitical influence for exporting nations (Huber, 2011). Consider President Putin's claim, although he denies ever making it, that Russia was an 'Energy Superpower', or President Trump's rallying cry to 'Drill, baby, drill' to promote US 'Energy Dominance'. Often, this perceived power proves illusory (Ashford, 2022). The geographically specific locations of fossil fuels have created complex global supply chains, which can be disrupted by transit states and at key maritime choke points for geopolitical advantage (Emmerson and Stevens, 2012; Palti-Guzman and Eyl-Mazzega, 2023; EIA, 2024a). International competition for access to fossil fuel supplies intensifies when markets are tight – that is, when supply struggles to meet demand. High prices tend to harm poorer economies the most, as they cannot absorb the additional costs. Similarly, producer economies experience negative impacts when fossil fuel prices remain low for extended periods. Sanctions on the oil and gas exports of producers are often wielded as a punitive tool of statecraft, especially by the US (Fishman, 2025). The effects of such measures are transmitted through the market, often leading to higher prices for consumers and/or lost revenue for exporters. As discussed below, access to and control over oil and its revenues have historically been sources of conflict. Many proponents of a rapid transition to renewable energy argue that it could eliminate the conflicts and repression linked to fossil fuels, offering a potential 'peace dividend'.

3.2 It's all about oil

A vast body of literature exists on the geopolitics of oil. For early seminal works, see Odell (1986) and Yergin (1990). For more recent treatments, see Colgan (2021), Ashford (2022), and Hanieh (2024). Most studies of oil start by recognizing that our modern world is, quite literally, made by oil. For well over a century, oil and its products have driven the global economy and now form the building blocks for the majority of the goods that we consume. As Yeomans (2004, xi) observed 'From the moment we wake up in the morning to the moment we go to sleep, oil controls our lives. Its influence reaches far into politics, international affairs, global economies, human rights, and the environmental health of our planet.' Not much has changed. Look around you and think about a world without oil. Notions such as 'Fossil Capital' (Malm, 2016), 'Carbon-Democracy' (Mitchell, 2011), and 'Carbon Capitalism' (Di Mizio, 2015) link the exploitation of fossil fuels to the development and reproduction of the capitalist economic system and global politics. This suggests that any attempt to reduce our reliance on fossil fuels in general, and oil in particular, will face significant inertia and difficulties, with profound economic and geopolitical consequences (Helm, 2017).

There can be no doubt about oil's strategic importance, nor the wealth and power it creates, but it is also linked to a range of 'pathologies'. As Huber (2011) has highlighted, there are conflicting narratives about oil, viewed on one side

Figure 3.1: Crude oil prices ($US) – the influence of geopolitical and economic events

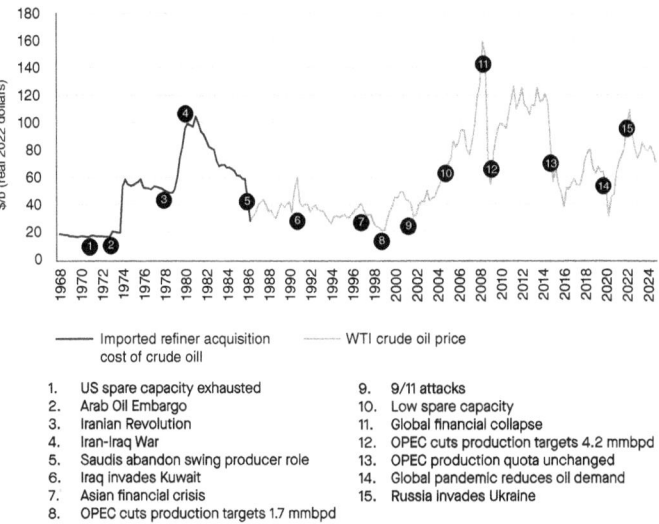

1. US spare capacity exhausted
2. Arab Oil Embargo
3. Iranian Revolution
4. Iran-Iraq War
5. Saudis abandon swing producer role
6. Iraq invades Kuwait
7. Asian financial crisis
8. OPEC cuts production targets 1.7 mmbpd
9. 9/11 attacks
10. Low spare capacity
11. Global financial collapse
12. OPEC cuts production targets 4.2 mmbpd
13. OPEC production quota unchanged
14. Global pandemic reduces oil demand
15. Russia invades Ukraine

Source: EIA (2025, np)

as 'powering progress' and driving economic growth, and on the other as the cause of conflict and repression. However, it is undeniable that fluctuations in oil prices greatly affect financial stability.

Figure 3.1 illustrates that various factors contribute to oil price volatility. Major geopolitical events are significant, but so are OPEC's actions to influence the market, either to push prices higher or to safeguard market share. Additionally, major economic shocks, such as the Asian Financial Crisis, the Global Financial Crisis, and the COVID-19 pandemic, also played a role. As a fungible product, crude oil prices are ultimately set by supply and demand; some events limit supply, while others reduce demand. The global nature of the market means that prices are similar worldwide, subject to local benchmarks, tax regimes, and subsidies. At the same time, oil's central role in the global economy means many factors influence its price at any given moment, making accurate predictions unlikely. However, it is notable that while Russia's invasion of Ukraine caused a short-term spike in oil prices in early 2022, prices have since remained relatively stable as the West pursued sanctions that did not fundamentally threaten Russia's oil supply to global markets. Moreover, Israel's war in Gaza has not, so far, significantly disturbed the oil market; in fact, OPEC+ has been trying to manage oversupply amid weak demand. Recently, fears of a global trade war triggered by US tariffs have further depressed oil prices, due to concerns over a possible global recession. A fundamental shift in the international oil market may be underway, and it is likely that a combination of economic and geopolitical pressures, rather than climate action, could lead to peak oil demand before the end of the decade.

3.2.1 Oil, conflict, and democracy

The literature on oil, war, and conflict mirrors its era – a period when the US remained a major producer but was progressively importing more oil. Following the events of the 1970s and the Carter Doctrine, which clarified US strategic interests in the Persian Gulf (Bridge and Le Billon, 2017, 135), the focus shifted towards the significance of oil in Cold War geopolitics and the US's efforts to preserve control and influence in the Middle East (Bromley, 2006). However, US interests did not always align with those of its European allies, who lacked substantial domestic resources (Thompson, 2022, 15–98). Additionally, oil wealth was often linked to repression, civil wars, conflict, and support for autocratic regimes. Ashford (2024, nd) calls this the 'Oil Security Paradigm', noting that 'energy security considerations were a core component of US–Soviet competition during the Cold War and shaped the US presence in the Middle East'. Scholars such as Klare (2001, 2004) and Rutledge (2005) argued that US foreign policy, as seen in the Gulf Wars of 1991 and 2003, was mainly driven by the need to control oil supplies to global markets. Since then, China's emergence as a competing centre of demand and influence has complicated these dynamics. Chinese demand played a key role in the resource Supercycle of the early 2000s, culminating in the global financial crisis (Jones and Steven, 2015; Thompson, 2022).

Subsequent events have once again reshaped the geopolitics of global oil, particularly as the US became a net oil exporter in 2020 for the first time since 1949 and is now the world's largest producer of oil and gas (EIA, 2024b). The US operates as a different kind of petrostate, consuming most of its production while relying on imports – mainly from Canada and Mexico – to supply heavier-grade oil and supplement domestic output. Consequently, in 2022, only 12% of US crude oil and petroleum imports came from the Persian Gulf. Nonetheless, as the hegemonic superpower, the US retains a strong strategic interest in the region to counterbalance China's growing influence. China's oil imports increased from nearly 9 million barrels per day (mbpd) in 2010 to 16 mbpd in 2023, with its share of global imports rising from 10% to 16% (IEA 2024a, WEO). The IEA (2024a) suggests that as China's demand for oil begins to decline, India may emerge as a key growth centre. Although the US's reliance on Persian Gulf imports has decreased significantly, events in that region and elsewhere continue to affect global oil prices, which in turn influence petrol prices in the US. This remains a significant political issue, emphasizing that physical independence does not assure price security.

In political science, a body of literature exists that examines the fortunes of oil-exporting states and the role that oil plays in causing conflict. Again, this work reflects a particular time of high oil prices and surging demand. This is part of a broader literature on 'resource wars' and 'resources and conflict' (Le Billon, 2001, 2007; Collier and Hoeffler, 2005). Given the importance of oil and its products to both economic activity and military operations, it has been regarded as a prize worth fighting for; however, the proceeds from oil exports also enable the financing of repression and conflict (Ashford, 2022, 239). Authors such as Jeff Colgan (2013) and Michael Ross (2001, 2004, 2012) led an academic endeavour to explore the relationship between oil and conflict. Their research revealed a complex relationship that is far from deterministic; there is nothing inevitable about the discovery of oil in a particular country leading to conflict. However, unfortunately, it has been the case all too often that oil wealth fails to deliver the prosperity it should bring (more on the economic dimensions in the next chapter).

Oil itself is not the problem; it is the fact that oil 'rents' (the difference between the value of crude oil production and the total costs of production) can promote conflict, erode democracy, or support autocracy. When a country becomes an oil producer and a net exporter, it is significant if a strong state already exists and safeguards are in place to manage the impact of oil rents. In that case, the new revenue stream can be effectively utilized to diversify the economy, and its potential adverse effects on the domestic political economy can be mitigated. However, if the state is weak and there are high levels of inequality within the country, then oil revenues can quickly become divisive, sometimes resulting in civil war. Equally, if the state is already autocratic, then oil revenues provide a means to maintain that autocracy. For example, the oil-exporting states in the Persian Gulf were not made autocratic by their oil wealth; however, that wealth has formed a particular type of 'rentier state', which is discussed in the next chapter. It is relatively rare for states to engage in conflict to gain control of another state's oil reserves, though it does happen. Access to oil was undoubtedly a key component of Nazi Germany's campaigns in the Soviet Union and North Africa during the Second World War, and a failure to secure oil played a significant role in its defeat (Kelanic, 2020, 92–114). There is also the Iraqi invasion of Kuwait that triggered the first Gulf

War in 1990. Most recently, there is the case of Venezuela's renewed territorial threats against Guyana following the discovery of large oil reserves in 2019 (Dodds, 2023). More generally, control of potential oil reserves is often a cause of territorial disputes over jurisdiction; for example, in the South China Sea, China is in conflict with all its neighbours over such claims. There are many other cases where states dispute the sovereignty of territory that may contain oil and gas reserves, but the prospects of dwindling demand and low prices may serve to diffuse such claims in the future.

Not surprisingly, a substantial body of literature focuses on the major oil-exporting states and regions, such as Kleveman (2003) on Central Asia, Rowell, Marriott, and Stockman (2005) on West Africa, and Clootens and Ben Ali (2021) on the Middle East and North Africa. The Soviet Union and Post-Soviet Russia have also been the focus of extensive and ongoing analysis. The consequences of declining oil and gas demand are significant for Russia's outlook and stability, which have important repercussions for the global geopolitical landscape and, therefore, merit further examination.

During the 1970s and 1980s, the export of West Siberia's oil and gas wealth to European markets laid the foundation for détente and improved East–West relations. The convertible currency earned through that trade supported a struggling Soviet economy through purchases of grain and Western technology. This was the focus of my PhD thesis in 1987. In the late 1980s, issues within the domestic oil industry and declining oil prices were key factors contributing to the collapse of the Soviet economy (Gustafson, 1989). A valuable lesson for economies overly reliant on fossil fuel rents is that even superpowers can fail. Following the collapse of the Soviet Union in 1991, several newly independent oil and gas exporting states, led by Russia and including Azerbaijan and Kazakhstan, attracted investment from Western oil companies eager to access new reserves. China also entered the scene, forging deals to build pipelines for importing Central Asian gas.

In Russia, while the gas industry remained state-controlled, several new private oil companies emerged and formed partnerships with international oil companies (Gustafson, 2012). In the early 2000s, Russia greatly benefited from high oil prices, allowing Vladimir Putin to consolidate his control over Russia's political economy (Miller, 2018). Control of the rents from oil, and less so gas, became a central element of the political economy of what Gaddy and Ickes (2005) called Russia's 'Virtual Economy'. Riding a broader wave of high prices and resource nationalism, in the early 2000s, Putin reasserted control over the oil industry, reducing the influence of both oligarchs and foreign oil companies, some of whom were compelled to relinquish control over their projects (Rutland, 2008; Bradshaw, 2009b). However, tying Russia's fiscal fortunes to hydrocarbon revenues also exposed it to oil price volatility (Ahrend, 2005; Conolly, Hanson, and Bradshaw, 2020). When oil prices have crashed, Russia has repeatedly attempted to pivot to China as a new pipeline export market, first for oil and then for natural gas. Again, Russia is a distinct type of petrostate, characterized by a diverse domestic economy but a federal government highly reliant on oil revenues (Bradshaw and Conolly, 2016; Vantansever, 2021). That weakness is now being exploited by the West's sanctions to penalize Russia for its invasion of Ukraine, though with dubious effectiveness. Russia has responded by reducing the flow of natural gas through its pipelines to European markets. As explained in detail below, this action has made clear the limitations of using natural gas for geopolitical purposes. Moving forward, Russia is particularly vulnerable to the energy transition and international climate action, perceiving both as threats to its national interests (Makarov, Chen, and Paltsev, 2020). In his most recent book on Russia and Climate Change, the West's leading expert on Russian energy, Thane Gustafson (2021, 1), observed that 'for Russia the consequences [of climate change] will be especially dramatic, for its economy and environment, and for its standing as a great power in the rest of the world'.

3.3 But natural gas is different

Just as work on energy security and geopolitics has neglected coal, it has also traditionally assumed that gas operates like oil. However, in recent years, a body of literature has emerged on the geopolitics of natural gas (Grigas, 2017; Bradshaw and Boersma, 2020). As noted earlier, most natural gas consumption still occurs in the country where it is produced; nonetheless, the share of natural gas that is internationally traded is increasing, and more countries are becoming involved in the industry. According to the International Gas Union (IGU, 2025, 10), in 2024, the total volume of LNG trade was approximately 411 million tonnes (560 billion cubic metres), involving 22 exporting countries and 48 importing markets. However, in 2024, natural gas accounted for 25% of the total energy supply, while LNG only accounted for around 11% of the total gas supply (EI, 2025, 37–42). For many decades, the global gas landscape was organized around regional markets, each with its pricing systems (Stern, 2012). Traditionally, the industry was dominated by long-term supply contracts between producers and consumers, with prices linked to oil; in essence, LNG supply chains became 'floating pipelines' with limited flexibility. This was driven by high capital costs and the need for secure demand to ensure a future revenue stream to fund the construction of LNG plants.

Historically, the global LNG industry was divided into the Atlantic and Transatlantic basins. Two main continental pipeline gas markets existed: North America and Eurasia, with the latter centred on pipeline exports between Russia and Europe. Today, the system is much more integrated, if not fully globalized (Bridge and Bradshaw, 2017). Short-term spot markets are expanding, with prices in Europe and Asia becoming increasingly correlated. The remarkable growth of

US LNG trade – enabled by the shale gas revolution – has been a key factor in global integration (Bros, 2012; Losz, Boersma, and Mitrova, 2019; Bradshaw and Boersma, 2020). Ironically, by mobilizing gas, promoting flexibility, and encouraging global market integration, LNG has made natural gas more like oil in terms of geopolitics (Harris, 2024). International pipelines create persistent interdependencies that can be exploited but are also vulnerable to the whims of transit states, making them susceptible to sabotage.

At sea, LNG tankers face threats akin to those experienced by oil tankers. For example, around 20% of global LNG exports – mainly from Qatar – pass through the Strait of Hormuz. Recently, transit through the Red Sea has been disrupted by the Houthis in Yemen (Sharples, 2024). All this indicates that the geopolitics of natural gas warrants analysis on its terms, especially as many in the industry regard it as a transition fuel likely to sustain significant demand growth through the 2030s. Although the idea of gas as a transition fuel is highly debated, this issue falls outside the scope of the current discussion (Kemfert et al, 2022). If gas is to play a larger role in the global energy mix, concerns about gas security and geopolitics are likely to become more urgent, not less. Likewise, concerns about the security and affordability of LNG may deter some countries from participating in the trade. However, for others, it provides a more diverse and adaptable alternative to pipeline gas imports.

3.3.1 A short history of Russia–Europe gas relations

Although LNG is gaining importance, the relationship between Russia and Europe regarding natural gas and geopolitics has always been, and remains, a central focus. Before analysing Europe's current gas crisis, it is necessary to explore history to understand how Europe became so dependent on energy imports from Russia. The pipeline gas trade between the Soviet Union and Europe began in 1968 with the first agreement with Austria (for detailed histories, see Högselius, 2012; Gustafson,

2020; Skalamera, 2023). In the following two decades, exports spread across Northwest Europe. In the context of the 1970s, the development of this gas supply network was seen as a counter to an increasingly unstable global oil market. The hope was that a strengthening economic interdependence would improve relations with Moscow. Therefore, this gas trade has always been a matter of geopolitics. The US never approved of Europe's increasing reliance on Soviet gas. Tensions escalated in the early 1980s following the Soviet Union's invasion of Afghanistan in December 1979 and the declaration of martial law in Poland in 1980. These developments led to US sanctions on oil and gas technology exports to the Soviet Union. At that time, Europe was involved in another gas-for-pipeline deal, and the issue even became a point of disagreement between US President Ronald Reagan and British Prime Minister Margaret Thatcher. Analysing events at the time, Jonathan Stern (1982, 36) concluded that 'However strongly President Reagan and his advisers may feel about the project, their actions have precipitated a disastrous episode – a classic example of how not to manage the alliance.' This view has contemporary resonance. From a gas security perspective, the Soviet Union did not have to worry about transit security, as the pipelines ran through its client states in Eastern Europe, members of the Warsaw Pact and Comecon. Furthermore, it had a vast resource base to draw on and long-term contracts provided it with security of demand, all the prerequisites for a secure and profitable relationship that benefited both East and West in Europe.

Things became much more complicated in 1989 when the Berlin Wall fell, followed by the collapse of the Soviet Union in 1991 and its disintegration into 15 newly independent states. The Russian Government retained control over the gas industry through its ownership of Gazprom, which still retains a monopoly over pipeline gas exports. But the geopolitical map of Central Europe was redrawn, and now a wall of transit states stood in the way of Russian gas exports

(Figure 3.3). Furthermore, many of those states joined the EU with its single gas market and competition rules. Despite all these changes, Russia could claim to be a reliable successor to the Soviet Union, and the gas continued to flow through the 1990s. However, problems began to arise in the transit of Russian gas through Ukraine, and Moscow increasingly utilized the gas trade to create issues for the Baltic States. Russian gas transit through Ukraine reached its peak at 141 billion cubic meters (bcm) in 1998 (Prokip, 2025). In 2004, the 'Orange Revolution' in Ukraine challenged Moscow's influence over the country. At that time, 80% of Russia's exports to Europe transited through Ukraine, and disagreements between the two nations' gas industries led to brief gas shortages in Central and Southern Europe during 2005–06 and 2008–09. This resulted in growing concerns about gas security, which the EU responded to by doubling down on its effort to build a more liquid and functional market (Sharples, 2013; Henderson and Pirani, 2014; Goldthau and Sitter, 2015). Some EU member states also sought to diversify their options by building LNG import terminals. Most significantly, in 2014, Lithuania installed the Floating Storage Regasification Unit (FSRU) *Independence,* which enabled LNG imports, thereby freeing it from the need to import Russian gas. Gazprom soon adjusted to the new reality in both the EU and the broader global gas market (Henderson and Moe, 2019). It also responded by acquiring infrastructure assets and establishing trading operations in the EU. However, Gazprom struggled to develop Russia's LNG potential, and the Kremlin liberalized the opportunity, allowing Novatek, a private company and Russia's second-largest gas producer, to become the dominant force in LNG exports to Europe (Henderson and Yermakov, 2024).

Gazprom invested billions of dollars in new pipelines to avoid transit risks in Ukraine (Figure 3.2). First, the Yamal–Europe pipeline, which runs through Belarus and Poland to Germany, opened in 2006. Then, in 2011, the Nord Stream 1 pipeline, extending from Russia through the Baltic Sea to Germany,

Figure 3.2: Europe's gas import infrastructure

Source: Breugel (2025, np)

began operations. Finally, the TurkStream pipeline through the Black Sea started operation in 2020. Agreement on the Nord Stream 2 pipeline was announced in 2015, just over a year after Russia's occupation of Crimea. It was supported by the German government, which led to divisions both within the EU and between Washington and Berlin (de Jong, Van de Graaff, and Haesebrouck, 2022). A modest degree of diversification was achieved in 2018 with the non-Russian pipeline route (TANAP), which links Europe to Azerbaijan. Nonetheless, between 2010 and 2020, Russia accounted for, on average, 38% of extra-EU gas imports (Statista, 2024). Europe continued to benefit from a plentiful supply of Russian pipeline gas, which set the price as the marginal (cheapest) source of supply into the EU's single market. The new transit pipelines rerouted gas away from Ukraine, where transit flows fell to 56 bcm by 2020 (Prokip, 2025).

Meanwhile, after an initial venture on Sakhalin with considerable foreign investment (Bradshaw, 2010), Moscow was considering a larger role in the global LNG market. It was also beginning to build a new pipeline to China, the Power of Siberia (Henderson, 2018). Throughout this period, Europe remained a modest and passive player in the international LNG market, avoiding long-term contracts as a way of securing supply and helping to balance the market when Asian demand was low.

3.3.2 Europe's global gas crisis

As the world emerged from the COVID-19 pandemic, energy and resource prices surged, as growing demand could not be met by supply. In late 2021, the situation worsened in Europe's gas market as Gazprom withdrew from the spot market, and prices began to rise (Figure 3.3). Initially, market analysts thought that Gazprom was prioritizing price over volume,

Figure 3.3: The Dutch Title Transfer Facility (TTF) gas price

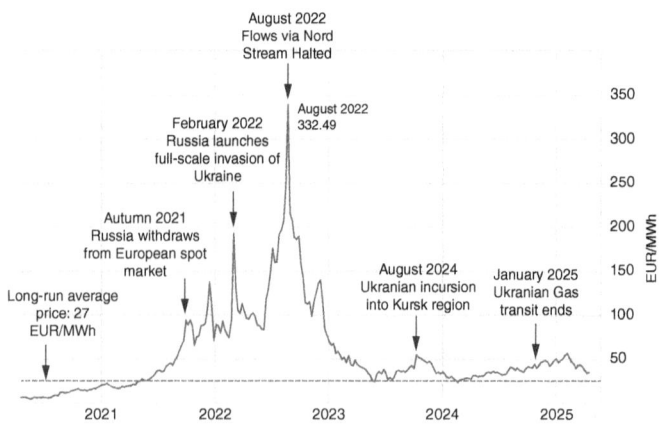

Source: Trading Economics (2025)

perhaps anticipating growing competition from surging US LNG exports. However, in the Spring of 2022, Moscow's motivations became clear when, on 24 February, Russia invaded Ukraine in a significant escalation of a simmering war that had started in 2014 with the invasion and annexation of Crimea (Henderson, 2024). The overall outcome has been a significant decline in Russian pipeline gas exports to Europe, not because the EU has imposed sanctions, as there are no direct sanctions in place, but due to Russian actions.

When President Putin insisted that payments for gas be made in Russian Rubles, in breach of existing contracts, many European counterparts refused. In response, Russia imposed sanctions that effectively shut down the Yamal–Europe pipeline. The Polish Government nationalized the section of the pipeline on its territory. Then, fearing a complete shutdown of Russian pipeline flows, European gas traders panicked as they sought to secure alternative supplies at any price, largely LNG, to fill Europe's gas storage ahead of winter. Brussels had required storage to be 90% full by November, which triggered a price war in summer 2022 with a global impact, as LNG prices reached unprecedented levels. Some emerging economies – for example, Pakistan – were priced out of the market. Some Asian buyers made a tidy profit re-exporting their cheaper oil-indexed LNG to Europe. Things became more difficult again when, on 26 September, a series of explosions damaged the Nord Stream pipelines on the Baltic seabed near the island of Bornholm in Denmark (de Jong, 2023). The damage affected both pipes of Nord Stream 1 and one of the two pipes of Nord Stream 2.

The second pipeline had only recently been completed, but the German Government had refused to certify it because of Russia's invasion of Ukraine. Nobody has taken responsibility, and it remains unclear who carried out this attack. Gradually, Russia's pipeline capacity was eroded, first with the Yamal–Europe pipeline (33 bcm), then Nord Stream 1 (55 bcm), with Nord Stream 2 (55 bcm) never entering operation (see

Figure 3.4: EU imports of Russian pipeline gas, Russian LNG, US LNG, and Total LNG

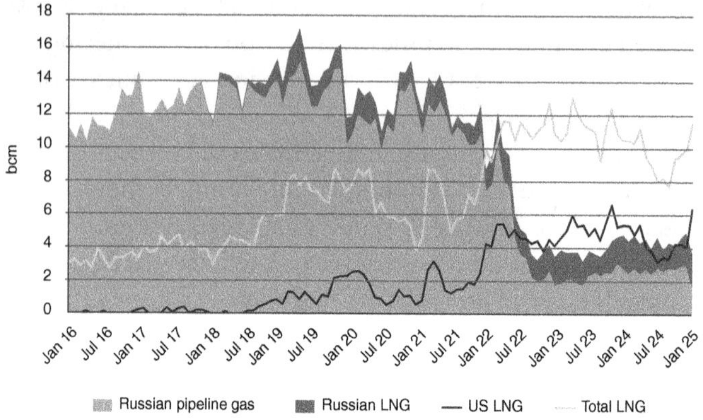

[Chart showing bcm on y-axis (0-18) from Jan 16 to Jan 25, with series: Russian pipeline gas, Russian LNG, US LNG, Total LNG]

Source: Corbeau (2025, np)

Figure 3.4). This left transit through Ukraine and TurkStream as the only remaining routes for Russian pipeline gas. In 2024, Russia exported 33 bcm through the TurkStream pipeline, with half of this volume consumed by Turkey, and 15.4 bcm through Ukraine, totalling 48.4 bcm. This compares to 174 bcm in 2021 (Sharples, 2025). Finally, at the end of 2024, the gas transit agreement between Russia and Ukraine expired and has not been renewed. As of early 2025, Greece, Hungary, and Slovakia continued to receive Russian pipeline gas via TurkStream. All this has resulted in a considerable loss of revenue for Gazprom. But how did Europe source alternative supplies?

The EU's initial response was its May 2022 REPowerEU programme (European Commission, 2025a,b), which aimed to save energy (particularly gas), diversify energy supplies (away from Russia), and produce more clean energy. Since then, significant progress has been achieved, with the EU's natural gas demand falling 18% between August 2022 and March 2024, resulting in 125 bcm less gas to be found, which exceeds the total US LNG production in 2024 (EI, 2025, 43). However, Europe's energy-intensive industries have suffered significant 'demand destruction' as high energy prices have forced them to shut down production, in some cases permanently. Meanwhile, households received hundreds of billions of euros in state support to cushion the impact of record price increases, but energy poverty still worsened. When it comes to diversifying energy imports, the EU has succeeded in replacing Russian gas supplies with imports from other sources. Norway has become the EU's leading supplier, followed by US LNG imports. The EU's LNG import capability was increased through the deployment of FSRUs. For example, the German Government provided significant funding to build LNG import capacity. The United Kingdom, no longer an EU member, utilized its significant LNG import infrastructure and interconnector pipelines as a bridge to help fill EU gas storage, recording record levels of LNG imports in 2023. According to IEEFA's (2025) European LNG tracker, in 2022, the EU imported 127.1 bcm of LNG at a total cost of €110.6 billion and 113.1 bcm of LNG for €61.6 billion in 2023. The US LNG industry and its gas traders have been the undoubted winners from Europe's crisis. In 2024, the US supplied the EU with 45% of its total LNG imports, and the EU purchased 43% of the US's total LNG exports (Corbeau, 2025). Although the US and EU have sanctioned Novatek's new Arctic-2 LNG plant, the EU has continued to import Russian LNG from plants on the Yamal Peninsula and the Baltic coast, although the US has now sanctioned the latter. Nonetheless, the share of Russian LNG in EU imports increased, and in the fourth quarter of 2024, it accounted for 22% of EU LNG imports, compared to 14% in the fourth quarter of 2023 (Eurostat, 2025). As of March 2025, the EU has placed sanctions on the transhipment of Russian LNG cargoes destined for third markets. Transhipment accounts for approximately 20% of Russian cargoes heading to the EU, and 38% of China's imports of Russian LNG are transhipped in EU ports. But one of the unintended consequences of the transhipment ban could be that even more Russian LNG is imported to the EU. The EU's reluctance to impose sanctions on Russian LNG reflected the tightness of the global LNG market at that time and its unwillingness to risk high prices.

While prices have decreased from the extremes of summer 2022, they remain well above pre-crisis levels (Figure 3.3), and Europe now relies on the global LNG market for its gas security. No longer a passive actor, it competes with Asia, particularly for spot LNG purchases. There has also been a significant element of good fortune, as mild winters in Europe and modest demand in Asia made it easier to fill storage before each winter. Consequently, storage levels remained higher than average at the end of the 2022–23 and 2023–24 winters. Things were more difficult in 2025 with a challenging winter in 2024–25 and the loss of transit through the Ukrainian pipeline system. The hope now is that an anticipated surge in new LNG supply will ease matters by 2026–27. Much of this increased LNG supply is expected to come from the US, which the EU has promised to purchase more of to alleviate trade tensions (Corbeau, 2025). In Europe, however, there is recognition that the only long-term solution is to reduce its reliance on imported fossil fuels, and it is hoped that the current energy crisis will accelerate the transition to low-carbon energy. According to EU data, in 2022, the EU generated more electricity from wind and solar than from gas, and in 2023, wind alone produced more electricity than gas (European Commission, 2024). At the same time, uncertainty about future gas demand means that European buyers remain unwilling to enter into long-term LNG supply contracts.

Both Russia and Europe enjoyed decades of trade in natural gas, but Vladimir Putin has destroyed that relationship (Sharples, 2025). However, Europe has paid a very high price for its new dependence on LNG. The current geopolitical environment under the Trump Administration injects further uncertainty into the future. The European Commission remains committed to ending all reliance on Russian energy imports by the end of 2027 and has published a roadmap; however, the details of how this will be delivered remain unclear (Yafimava, Ason, and Fulwood, 2025). Some in Europe, such as Hungary and Slovakia, advocate for restoring Russian gas imports via Ukraine (Vantansever and Goldthau, 2025). There is even discussion about using the remaining intact section of the Nord Stream 2 pipeline to increase Russian gas flows, in the hope of lowering energy prices in Europe. This would benefit Russia, which has been struggling to persuade China to agree to a second Power of Siberia pipeline. China is well aware of the risks associated with excessive reliance on Russian gas imports (Mitrova, 2023). Analysis that I have worked on indicates that Russian gas exports could decrease by 31–47% by 2040 if new markets remain limited, or by 13–38% if a strategic pivot to Asia – which includes building the second pipeline – is implemented (Pye et al, 2025). At a summit meeting in Beijing in September 2025, Russia announced that a legally binding memorandum had been signed for the Power of Siberia 2 pipeline; however, an agreement still needs to be reached on the price of the gas, commercial terms, and the timetable for completion (Corbeau, Downs, and Mitrova, 2025).

The pace of progress on climate change mitigation also plays a significant role, influencing future global demand for natural gas. Vladimir Putin's attempt to weaponize Europe's dependency on Russian gas has proved to be a pyrrhic victory. In a broader, longer-term context, Europe's gas crisis challenges the notion that LNG exports are secure and affordable – even for some of the world's wealthiest economies – which poses problems for an industry that anticipates growth in the 2030s, particularly in price-sensitive emerging markets in Asia. This has been further complicated by President Trump's willingness to use LNG as a tool of his statecraft. It also reminds nations that energy security issues caused by the geopolitics of fossil fuels remain an ongoing threat. Europe relied on Russian pipeline gas during a period of geopolitical instability in the global oil market. Today, despite broader geopolitical uncertainties, the oil market appears less volatile compared to the natural gas market. The key lesson here is that geopolitical conflict and fossil fuels are closely connected. Therefore, for many countries that import fossil fuels, the solution lies in accelerating the shift to low-carbon energy sources to reduce dependence on fossil-fuel-related geopolitical risks.

3.4 Conclusions: Energy system transformation and the geopolitics of oil and gas

This chapter has explored the complex relationship between oil, gas, and geopolitics. The importance placed on the production and trade of these fossil fuels has strengthened the position of major exporters. Simultaneously, oil remains closely linked to international conflict, and the oil market is marked by volatility. Until recently, natural gas was seen as relatively more secure than oil, although concerns about dependencies created by pipelines continued. The expansion of the global seaborne LNG trade has made gas more like oil. Recently, Russia's weaponization of natural gas has brought about significant changes in the international gas trade. For most states that are net importers of oil and gas, the priority has been to secure sufficient supplies to meet demand at affordable prices, thereby benefiting consumers and supporting economic growth. This shapes our current understanding of energy security. However, the situation will evolve, first for oil and then for gas, as climate action transforms the global energy system and permanently reduces demand for fossil fuels. As previously explained, the speed at which this will happen remains a critical uncertainty. A rapid and disorderly high-carbon transition risks market disruption, leading to economic and geopolitical instability in exporting states; the characteristics of these states are discussed in the next chapter.

FOUR

The Resource Curse, Rentier States, and Unburnable Carbon

This chapter explores the consequences, both domestically (internal stability) and internationally (external relations), of declining oil and gas demand for countries that depend on revenue from oil and gas exports. It begins by analysing the fortunes of exporting states to understand the potential effects of falling export income. The second section then turns to a specific class of oil- and gas-exporting states, the so-called 'rentier states' of the Persian Gulf, and the measures taken to prepare for a future without oil. The final section addresses the issue of whose fossil fuels are most likely to remain in the ground as energy system transformation accelerates. The conclusions then explore the geopolitical repercussions of phasing out fossil fuels.

4.1 The 'resource curse' revisited

On the face of it, one would think that being a fossil fuel exporter would give a state an advantage over those that have to import large quantities of fossil fuels; however, the evidence suggests that such 'resource-abundant' economies often perform worse than resource-poor economies. Since the 1990s, a body of literature has emerged that argues for a 'resource curse', suggesting that resource wealth may be more of a curse than a blessing (Auty, 1993; Rosser, 2006; Frankel, 2010; Alssadek and Benhin, 2023). However, similar to the oil and conflict literature discussed in the previous chapter, the resource curse is not inevitable, and not all resource-rich states have experienced it. It appears that oil-exporting states are more vulnerable, mainly due to the size of the revenues involved. This section aims to answer three questions: first, who are the so-called 'producer economies', those heavily reliant on revenue from oil and gas exports; second, what are the characteristics, causes, and consequences of the resource curse; and third, how might we rethink the resource curse in the context of the high carbon transition and declining demand for fossil fuels?

4.1.1 Who are the producer economies?

There are various ways of defining what the IEA (2018, 11) calls 'producer economies', defined as 'large oil and gas exporters who are pillars of global supply and rely on hydrocarbon revenues to finance a significant proportion of their national budgets'. Unfortunately, there seems to be no agreement on what defines a 'large producer' or what constitutes a 'significant' proportion of a nation's budget. Regardless of the measures employed, it is clear that producer economies form a heterogeneous group with varying levels of dependence and importance to the global energy geopolitics.

Ashford (2022, 19–25) proposes that there are three distinct categories of petrostates: oil-dependent states, oil-wealthy states, and superproducer states. Following Ashford, oil-dependent states rely heavily on oil revenues, with dependence defined as exceeding 10% of GDP. Figure 4.1 presents data from the World Bank and employs a slightly different measure that calculates oil rent (the difference between the value of crude oil production at regional prices and total production costs) as a percentage of GDP, which generally highlights the more extreme cases of dependence.

This list of oil-dependent states comprises many nations synonymous with the pathologies of oil production and

Figure 4.1: Countries where 'oil rent' is more than 10% of GDP, 2021

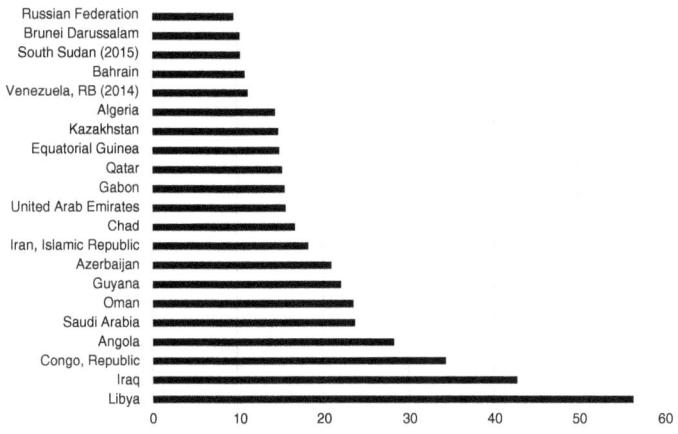

Source: World Bank Development Indicators Database. https://data.worldbank.org/indicator/NY.GDP.PETR.RT.ZS?locations=1W

plagued by conflict and instability. Ashford (2022, 20), using a slightly different indicator, observes that two thirds of her sample are nondemocratic, mainly situated in the Middle East and Africa, and are also largely economically underdeveloped. She notes that such oil-dependent states are often characterized by 'institutional weakness, corruption, or a state-controlled economy.'

Ashford's second group, oil-wealthy states, are enriched by having oil wealth, without being necessarily dependent on it, often because they have large, diversified economies. In these economies, the per capita measure of oil wealth may be more appropriate. Figure 4.2 offers an approximate measure based on oil production per capita; this measure reflects the physical volume of production rather than the wealth it creates. While there is significant overlap with the first group, this group includes large, diverse economies such as Canada and the United States, as well as major producers with small populations, including Norway. Additionally, note Guyana's position following its recent oil discoveries. In this category,

Figure 4.2: Oil production per capita, 2023

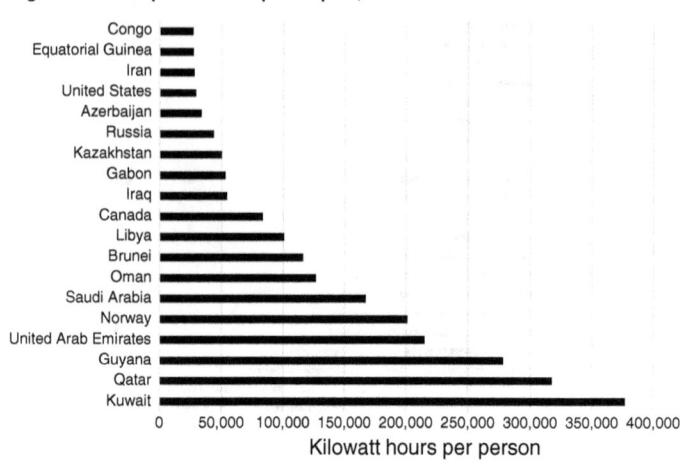

Source: *Our World in Data*, https://ourworldindata.org/grapher/oil-prod-per-capita

two thirds of oil-rich states are nondemocratic, most are minor powers, and over half are located in the Middle East. Ashford's final group includes the superproducer and superexporter states. According to Ashford (2022, 24), these are 'countries that occupy a unique place in international affairs because of their producer status'. Figure 4.3 shows those countries that account for more than 2% of global oil production.

Table 4.1 lists the most significant traders of crude oil and oil products; not all of them are superexporters or major crude oil producers. However, the top 6 account for 63% of global exports. Some countries on the list could be seen as 'supertraders' that import crude oil produced elsewhere and turn it into refined products, for both local use and export. Their presence shows that not only producers and exporters of oil face the transition risk from declining oil demand. Countries such as Singapore and India, which have invested in refining crude oil and exporting its products, are also vulnerable. Referring to Ashford's typology, the superproducers and superexporters

Figure 4.3: Major oil producing states, 2023

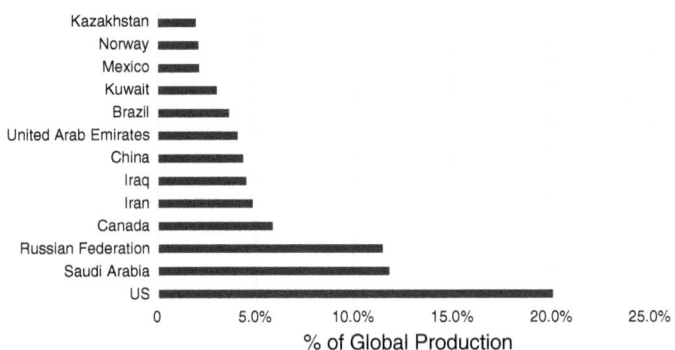

Source: Energy Institute (2024, 21)

include some of the world's largest economies, many of which are large, diversified, and democratic. However, there are also a few autocratic states that leverage their oil wealth and market power for geopolitical aims, most notably Russia.

When it comes to exposure to the low-carbon transition, a group of smaller, less developed, and less diversified energy-dependent states is the most vulnerable (Sarka and de Waal, 2024). The think-tank Carbon Tracker Initiative has conducted extensive research into the potential fiscal impact of the energy transition on 40 petrostates to identify those most exposed to transition risk (Prince et al, 2023). They used the IEA's Announced Pledges Scenario (see Chapter 2), which results in what they call 'moderate demand' for oil and gas, to stress test the impact on total government revenues. Table 4.2 shows the countries classified into five tiers, from the least vulnerable to the most vulnerable.

By comparing the amount of oil and gas revenue at risk in a moderate demand scenario with the dependence on oil and gas revenue as a proportion of government income, their analysis concludes that 28 out of 40 petrostates would lose more than half of their revenue, and that $8 trillion in expected revenue would be lost between 2023 and 2040. Their analysis also

Table 4.1: Major traders of crude oil and oil products, 2023
(Millions tons and %)

	Crude exports	% of Global exports		Product exports	% of Global exports
Saudi Arabia	349.1	16.4	US	258.0	21.2
Russia	240.8	11.3	India	93.3	7.7
Canada	207.2	9.7	UAE	92.5	7.6
US	185.0	8.7	Russia	90.9	7.5
Iraq	184.2	8.7	Singapore	78.0	6.4
UAE	170.7	8.0	China	61.5	5.1
Kuwait	81.1	3.8	Saudi Arabia	60.8	5.0
Mexico	53.8	2.5	Kuwait	37.4	3.1
Singapore	2.0	0.1	Canada	32.3	2.7
India	1.9	0.1	Iraq	19.8	1.6
China	1.2	0.1	Japan	12.5	1.0
Japan	neg.	neg.	Mexico	9.2	0.8

Source: Energy Institute (2024, 33–5)

Table 4.2: Vulnerability of petrostates

Least vulnerable				Most vulnerable
Tier 1	Tier 2	Tier 3	Tier 4	Tier 5
Cameroon	Bahrain	Bolivia	Algeria	Angola
Colombia	Egypt	Ecuador	Azerbaijan	Chad
Myanmar	Iran	Iraq	Brunei	Congo
Suriname	Mexico	Kuwait	Kazakhstan	Equatorial Guinea
Ukraine	Papua New Guinea	Libya	Oman	Gabon
Yemen	Uzbekistan	Malaysia	Saudi Arabia	Nigeria
		Norway	South Sudan	Qatar
		Russia	Sudan	Timor-Leste
			Trinidad and Tobago	Venezuela
			UAE	

Source: adapted from Prince et al (2023, 2)

identified several emerging petrostates or new producers – Guyana, Mauritania, Mozambique, Senegal, and Uganda – that could soon face transition risks if their revenues are not adequately managed and whose prosperity might be short-lived (Marcel et al, 2023). A somewhat similar analysis by Van de Graff (2023, 186–7) compares the level of exposure to the level of resilience. It identifies a group of countries that are most at risk, including Iraq and Libya, Venezuela, and Nigeria. The IMF has also identified a group of emerging and developing economies with higher fossil fuel dependency that will be significantly impacted by falling oil demand (Puyo et al, 2024). Not all petrostates will be affected to the same extent (Goldthau and Westphal, 2019). However, some of the most significant producer economies are in the higher tiers, and there is also a concentration of African and Middle Eastern states. Many of the most exposed states appear on the list of 'losers' identified by the GeGaLo index, which measures the geopolitical gains and losses of 156 countries resulting from a full-scale transition to renewables (Overland et al, 2019). The following section explains why those most dependent are the least equipped to handle a disorderly transition away from fossil fuels.

4.2.1 Characteristics, causes, and consequences of the resource curse

There is a substantial body of work that analyses the relationship between resource wealth and economic underperformance. The seminal work on this issue is a series of papers by Sachs and Warner (2001), published in the 1990s and summarized in a 2001 paper. They present a 'crowding out' hypothesis, in which the resource sector distorts the economy by crowding out activities, particularly manufacturing, that drive growth in nonresource-abundant economies. They concluded that 'Almost without exception, the resource-abundant countries have stagnated in economic growth since the 1970s, inspiring the term "curse of natural resources". Empirical studies have shown that this curse is a reasonably solid fact' (Sachs and Warner, 2001, 837). In a later review, Frankel (2010, 2) suggested that 'The evidence for a clear-cut negative correlation is not very strong, masking as many resource successes as failures. But it certainly suggests that there is no positive correlation between natural resource wealth and economic growth.' There is also evidence that learning has helped reduce the resource curse issue in recent decades (Auty and Furlonge, 2019, 21–3). Auty and Gelb (2001), using Auty's notion of the 'Staple Trap' – which drew on the work of Canadian economic historian Harold Innis from the 1960s, further developed by Mel Watkins (Stanford, 2020) – concluded that 'The economy is ... deflected from its comparative advantage and accumulates economic distortions that retard diversification and/or cause the economy to regress into a staple trap of dependence on the weakening primary sector.' The idea of 'Dutch Disease', which stems from the Dutch economy's experience after discovering natural gas, proposes that the appreciation of the domestic currency that followed harmed the local manufacturing sector, making it unable to compete with cheaper imports (Corden, 1984; Alssadek and Benhin, 2024). Thus, various economic mechanisms render the domestic economy overly dependent on the resource sector, leaving it vulnerable to the boom-and-bust cycles of price volatility. As we have seen, the oil price has been particularly volatile, and this volatility has intensified in recent years. The general advice for addressing this situation is to diversify the economy by incorporating manufacturing and services to enhance economic resilience and provide an alternative revenue stream for the government. However, the scale of the challenge varies significantly, and it is often easier said than done, depending heavily on the capacity of the state (Karl, 1997; Jones Luong and Weinthal, 2010; Alsharif, Bhattacharyya, and Maurizio, 2017; Blankenship et al, 2024).

Building on the earlier discussion of oil, democracy, and conflict, many oil-exporting countries are autocratic, and control over oil rents is highly centralized. Furthermore, the oil sector tends to be capital-intensive rather than labour-intensive and does not generate many well-paid jobs; instead, it is the ruling elites who benefit, often through bribery and corruption (Mahdavi, 2020; Sakar and de Waal, 2024). Similarly, significant resource developments can become enclaves with limited linkage to the host economy. The overall outcome is growing social inequality, which often has a geographical aspect, with oil-producing areas not necessarily gaining from the wealth beneath their feet, while the national capital benefits from their resources. The situation in Nigeria's Niger Delta is a case in point (Watts, 2004). All too often, those oil-producing regions also suffer from significant environmental degradation, which erodes living standards and can be a cause of conflict (Bamidele and Erameh, 2023). The issue of centralized control is often worsened by a shortage of human capital and the poor decision-making of the ruling elite, which Karl (1999) called 'Petromania'. Instead of investing in education to enhance human capital or adopting a strategy of economic diversification to reduce reliance on volatile oil prices, there is a tendency to fund rent-seeking initiatives, such as vanity projects – cathedrals in the desert – that consume large amounts of capital, resulting in a significant opportunity cost for the economy. At the same time, a lot of money is spent on armaments, a factor that increases the likelihood of conflict (Do, 2021). For example, countries in the Gulf allocate, on average, 4–5% of their GDP on defence, which is well above the global average of 2% (McGerty, 2022). Finally, as explained in more detail in the next section, in a resource-rich economy, there is no direct link between effort and reward, and the dominance of the resource sector can hinder development, investment, and entrepreneurial activity in the rest of the economy, leading to a 'failure of productivity'. Domestic energy prices are often heavily subsidized, resulting in high levels of domestic energy consumption that can erode the exportable surplus. The IEA (2018, 59) noted that in many producer economies, the cost of domestic electricity prices for residential consumers did not cover the cost of supply, removing any commercial incentive to expand supply if the sector is in private hands. Hence, in Nigeria, a major oil-exporting country, in 2022, nearly 40% of the population still lacked access to electricity (World Bank Development Indicators, 2025).

4.2.2 The resource curse and the high-carbon transition

In more extreme cases, all the problems discussed above create a situation where the very nature of the political economy of the oil-exporting state prevents it from adopting policies that would reduce its dependence on resource rents and exposure to price volatility; instead, it becomes addicted to rent. This leaves these states massively exposed to the consequences of falling fossil fuel demand and low prices; however, there are well-understood policy prescriptions to address the resource curse (Humphreys, Sachs, and Stiglitz, 2007). Stevens et al (2015, 18–19) suggested a range of policy imperatives that amount to 'good governance' (Box 4.1).

Box 4.1: Policy prescription to address the resource curse

- Establishing strong laws and institutions.
- Using transparent and competitive contracting processes.
- Smoothing revenue volatility by using sovereign wealth funds and/or future generations' funds.
- Spending revenues on long-term public infrastructure and debt repayment.
- Ensuring the transparency of revenues and spending from the resource sector.
- Increasing accountability and democratic participation.
- Minimizing and compensating for the environmental impacts of resource extraction.
- Strengthening linkages between extractive industries and the local economy.

Much like conducting your own financial health check, producer economies, both old and new, should carefully assess their current situation and reflect on the potential impacts of a significant decline in resource revenues in the coming decades. They could employ scenario planning, along with an assessment of different levels of physical and transition risks, to explore a range of possible future outcomes. Bradley, Lahn and Pye (2018, 30) conducted such an analysis and considered the 'resource curse through a carbon lens'. Their work contributes to a modest body of literature rethinking the resource curse within the context of the high-carbon energy transition (Lahn and Bradley, 2016; Prince et al, 2023; Saha et al, 2023). Given the discussion above, it is clear that much more needs to be done to understand the risks facing what Sarkar and de Wall (2024, 321) call 'fragile fossil-fuel producing states' in the face of falling demand. Van de Graff (2023, 190–1) presents a 'Petrostate Toolkit' with four options: a race to sell oil, restraint of production, hedging (diversification into other parts of the fossil fuel supply chain), and adapting (diversification into other sectors of the economy). These strategies are not mutually exclusive, and some states are adopting a mixed approach; however, many still appear to be in denial. Even worse, some actively block the efforts at COP to agree on phasing out fossil fuels, providing clear proof of their 'rent addiction' and a refusal to consider a future without fossil fuels.

4.3 The Gulf Rentier States and declining oil rents

This section aims to accomplish two objectives: first, it explains the origins and nature of rentier state theory and its evolution; second, it reviews the evidence for economic diversification among the oil- and gas-exporting member states of the Gulf Cooperation Council (GCC). The Gulf States are identified for four reasons. First, as shown above, most are highly reliant on oil and/or gas revenues; second, they are major oil exporters; third, they have clear strategies to adapt to a future with limited hydrocarbon income; and fourth, the region they occupy has been a centre of geopolitical conflict over hydrocarbons.

Therefore, failure to address the political, economic, and social effects of the high-carbon energy transition in the GCC countries will have significant repercussions for the global energy system and global geopolitical stability (Mills, 2020; Fattouh and Sen, 2021).

4.3.1 The Rentier State model

The 'Rentier State' is a political economy model that serves as a theoretical and analytical framework for examining and understanding the economic, political, and social structures that have developed around the management of oil revenues (and gas in Qatar's case) in Middle Eastern and North African (MENA) countries. Here, the focus is on the Gulf States, which exhibit significant differences as a group (Moritz, 2020). North Africa, Iran, and Iraq present a different case, and they are even more vulnerable to the risks associated with the high-carbon transition (Aoun, 2013).

The origins of Rentier State Theory can be traced back to a study by Mahdavy (1970) on pre-revolutionary Iran. It was then further developed by Beblawi (1987) and Luciani (1987) in an edited volume that explored the consequences of the oil boom for oil-exporting states in the MENA region. Although there are clear overlaps with the resource curse, the term Rentier State has become synonymous with the Monarchies of the GCC. According to Beblawi (1987, 51), a rentier state relies on substantial external rent, in this case, the rents obtained from the export of oil and gas. Furthermore, the state relies on this external rent due to the lack of a strong, productive domestic economy. Within the state, only a few are involved in generating this rent, while the majority participate in its distribution and consumption. In terms of the few, it is the government of the rentier state that is the principal recipient of the rent and is responsible for how it is used. Luciani (1987) expanded on Beblawi's framework, distinguishing between 'production states' and 'allocation states'. Thus, the purpose of the rentier states of the GCC is to capture, manage and allocate the rents earned from the export of oil and gas. Their National Oil Companies (NOCs) play a pivotal role in this process, both in producing rent and benefiting from it. As Krane (2019) explained, rentierism has a long history in the Gulf, and it is for that reason that the ruling Gulf Monarchies were able to capture the rents associated with hydrocarbon exports.

Furthermore, Beblawi (1987, 52) maintained that there is a 'rentier mentality' that is the result of a break in the 'risk-reward causation'. For citizens in the rentier state, reward (wealth) is not related to their work or risk bearing, but rather to chance or situation. Luciani (1987) explained that in rentier states, rents are distributed based on political criteria – such as loyalty, proximity to the ruling elite, and family ties – which reinforces traditional loyalties. This has translated into a 'social contract' or 'rentier bargain' between the ruling elite and the citizenry, whereby the state provides for their material needs – often in the form of a government job – in return for their acceptance of the political order (Moritz, 2020, 164). In such a context, the absence of taxes on citizens and the prevalence of subsidies guarantee an acceptable living standard, resulting in the phrase 'no taxation, but no representation' (Krane, 2019). It also results in a weak bureaucratic capacity, which is linked to the resource curse and the role of institutions in managing risk.

A great deal has changed in the GCC since the concept of the rentier state was first introduced. There have been considerable advances in the level of economic development and the living standards of the citizens in the region. The glass towers on the shores of the Gulf have been built by an army of low-skilled immigrant workers, who now also support the needs of the citizens and a class of technocratic expats that have helped to expand the oil and gas industry in the region, as well as staff its rapidly growing service and knowledge economy. New islands have been constructed, and globe-spanning airlines have transformed the Gulf into a hub for corporate offices, a key point on the Formula 1 calendar, a popular holiday spot, and a convenient stopover between Europe and Asia. Research on the political economy of the Gulf suggests that 'rentierism' has evolved with the times and has entered what Gray (2011) called an era of 'late rentierism'. However, while there are significant differences between the states of the GCC, the essential elements of the rentier state remain intact (Beck and Richter, 2021).

While the region has experienced the full impact of globalization and economic growth, along with some easing of social norms, there has been little in terms of democratization, nor has much been done, until recently, to boost female participation in the workforce (Ross, 2012, 231). The Gulf Monarchies remain autocratic, and at times, social unrest – most notably in the aftermath of the Arab Spring of 2010–11 – has been offset by economic benefits provided to their citizens. Unfortunately, human rights abuse and repression remain a force at the disposal of the ruling elite. There is also evidence that the 'rentier mentality' remains in place (Hertog, 2020); however, it has become increasingly apparent that the current model is no longer sustainable (Davidson, 2012). Because it is based on the export of nonrenewable hydrocarbons, it was inevitable that it would come to an end at some point in the future. However, two factors are bringing that future forward earlier than expected. First, there is the high-carbon transition itself, which necessitates a reduction in hydrocarbon consumption. This promises increased volatility and lower prices but ultimately results in the permanent loss of the very rents that rentier states rely on. Secondly, and even more urgently, is the pressure caused by the demographic situation in the Gulf States, especially in Saudi Arabia. The region's oil boom has been associated with rapid rates of population growth and substantial investments in healthcare and education, which have now translated into a significant number of well-educated men and women entering the labour market, expecting their share of the rentier bargain (Hertog, 2025). This comes at a time when the ability of some rentier states – notably Saudi

Arabia – to meet their obligations is being challenged (Hertog, 2024; Luciani, 2024). This is primarily due to their limited success in creating a thriving private sector that is independent of Government support.

4.3.2 Evidence of diversification

The six states of the GCC are a diverse group in terms of history, territory, population, and rentier wealth. Herb (2014, 2) has distinguished between the extreme rentiers – Kuwait, the UAE, and Qatar – on the one hand, which enjoy the highest per capita rent incomes in the world, and the not-so-rich, middling rentiers – Saudi Arabia, Oman, and Bahrain – on the other hand. The most affluent rentiers have enough wealth to maintain their rentier model for some time to come; the second group faces challenges that make diversification more urgent. The issue for Bahrain and Oman relates to their declining oil output, whereas the problem for Saudi Arabia, as mentioned above, is the large number of young Saudis entering the labour market.

When evaluating diversification, the main challenge is how to measure the reliance on oil rents in Gulf economies and the progress made in diversifying their economies. Table 4.3 shows data collected by the World Bank on the proportion of oil rents in GDP from 1990 to 2021. Several points can be drawn from these figures. First, there is a general trend of decreasing dependence on oil rents across the region. Second, the data for 2014–15 highlights the significant impact that the drop in oil prices at that time had on the region's economy. This prompted renewed efforts to diversify. Third, there are notable differences among the countries in the GCC. As mentioned above, in Oman and Bahrain, the declining role of oil rents reflects the depletion of their reserves and falling production. However, in Oman, this has been partly offset by expanding gas exports. The situation in Qatar is also understated, as it relies heavily on LNG exports rather than

Table 4.3: Oil rents as a percentage of GDP, GCC States, 1990–2021

	1990	2000	2014	2015	2020	2021
UAE	35.3	20.8	22.9	13.1	10.5	15.7
Bahrain	37.0	18.7	18.9	9.8	6.7	10.9
Kuwait	38.9	50.3	53.2	36.4	38.7	27.6
Oman	49.6	44.5	31.7	18.9	15.0	23.5
Qatar	48.7	40.8	21.5	13.0	10.6	15.3
Saudi Arabia	46.4	41.7	40.3	24.0	16.0	23.7

Source: World Bank Development Indicators Database, https://databank.worldbank.org/source/adjusted-net-savings/Series/NY.GDP.PETR.RT.ZS

oil. The UAE includes Dubai and Abu Dhabi, with the latter being a significant oil exporter. The continued reliance of Kuwait and Saudi Arabia is apparent, although considerable progress has been achieved.

Figure 4.4 presents a different perspective on rent dependence, illustrating how the share of hydrocarbons (both oil and gas) in government revenue fluctuates. Once again, there is a general trend of decreasing reliance over the decade from 2013 to 2023; however, dependence remains very high. This is particularly significant because the rentier state model relies on government expenditure to redistribute the rents.

The fiscal breakeven price (sometimes called the 'social cost of oil') is the minimum price of oil that a government needs to balance its budget (Table 4.4). The numbers present a mixed picture, and the problem states are those most dependent on oil revenues for government spending – specifically, Bahrain, Kuwait, and Saudi Arabia, with the latter being the most concerning due to its significant role in the global oil market and its geopolitical weight.

Overall, the diagnosis is complex, and a substantial body of literature has emerged on the political economy of the

Figure 4.4: GCC states – hydrocarbons' share of government revenues, 2013 and 2023

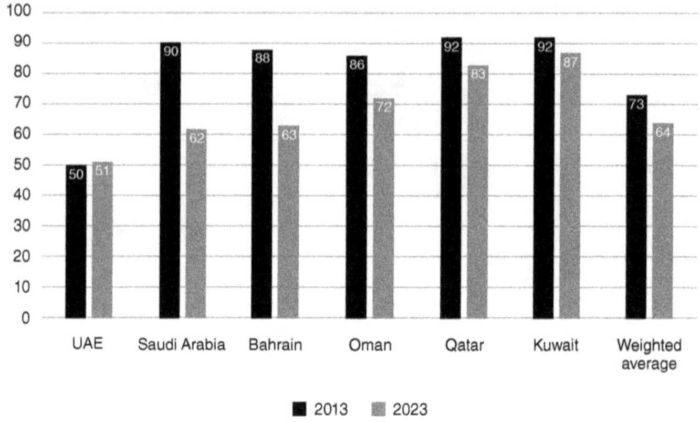

Country	2013	2023
UAE	50	51
Saudi Arabia	90	62
Bahrain	88	63
Oman	86	72
Qatar	92	83
Kuwait	92	87
Weighted average	73	64

Source: Richards (2024, np)

Gulf, the ongoing importance of rentierism, and its future prospects (Hanieh, 2018). In a similar vein to Van de Graff's (2023) 'Petrostate Toolkit', Krane (2019, 95) suggested that the GCC states have four options for the future: first, invest in the upstream to increase oil and gas production; second, diversify the energy mix by investing in nuclear and renewables; third, diversify their economies beyond hydrocarbons; and fourth, attack demand. Clearly, these options are not mutually exclusive. The fourth option – 'attack demand' – highlights how energy price subsidies in the GCC states have led to some of the highest per capita energy use in the world, to an extent that domestic demand is eroding the amount of oil available for export. Across the region, measures are now being implemented to cut subsidies, creating a win–win–win situation that decreases the cost to the government, increases the exportable surplus, and helps to cut GHG emissions. This motivation differs significantly from the discussion in Chapter 1 about energy efficiency and demand reduction; here, the primary

Table 4.4: GCC, fiscal breakeven oil price

	Average 2000–19	2020	2021	2022	2023	2024
Bahrain	83.2	120.6	131.6	133.6	126.2	129.1
Kuwait[1]	n/a	63.4	87.0	63.2	70.7	66.3
Oman	68.7	86.4	76.7	62.1	72.2	66.4
Qatar	45.1	49.3	47.9	44.7	44.8	41.5
Saudi Arabia	80.4	76.3	83.6	85.82	80.87	75.07
United Arab Emirates	50.0	51.7	53.0	46.6	52.8	53.9
Average Brent Crude Spot Oil Price ($/bbl)	64.59	41.84	70.91	101.32	82.64	81.00

[1] Kuwait's fiscal breakeven oil price is calculated using the fiscal balance before the 10% revenue transfer to the Future Generations Fund and includes investment income.

Note: Data for 2000–19, 2020, and 2021 are from the IMF. Data for 2022, 2023, and 2024 are from the World Bank's Prosperity Data360. The oil price data are from the Energy Institute, except for 2024, which comes from the EIA.
Source: IMF (2024), World Bank Group (2025), Energy Institute (2024, 30) and EIA (2025)

objective is to conserve oil for export. When it comes to broader economic strategy, Roberts (2023) identifies four strategies: the promotion of long-term visions, such as Saudi Arabia's Vision 2030 (Hertog, 2024); the rule of megaprojects (such as Saudi Arabia's NEOM); the desire to create knowledge economies; and the foundation of Sovereign Wealth Funds (SWFs). The latter holds global importance, as the GCC now possesses significant assets worldwide, serving as a hedge against declining oil prices and a fund for future generations. Abdel-Fattah (2023) states that the GCC SWFs hold assets worth $4.8 trillion, accounting for 40% of the world's total wealth held in SWFs.

There is much more that could be said about the role of hydrocarbons in the Gulf and their global impact, but a final thought must go to the prospects for the region, and for

Table 4.5: World oil production and price by IEA's 2024 WEO scenarios

	2023	STEPS 2030	STEPS 2035	STEPS 2050	APS 2030	APS 2035	APS 2050	NZE 2030	NZE 2035	NZE 2050
World Oil Production (mb/d)	96.9	99.2	96.5	90.3	90.4	79.9	52.1	76	55.9	22.4
OPEC Share (%)	34	33	34	40	34	36	41	35	39	51
Price (USD [2023]/barrel)	82	79	78	75	72	67	58	42	33	25

Source IEA (2024b, 137)

producer economies more generally. The price projections associated with the IEA's (2024a) energy scenarios illustrate the level of uncertainty about future oil prices (Table 4.5). As one might expect, a faster transition results in lower oil production and lower oil prices. Interestingly, in all cases, OPEC's share of future production increases, and the reasons for this are explained below. In their middling 'APS' case, it is clear that unless the GCC states reduce their reliance on oil rents, they will face significant financial problems in the 2030s that threaten the 'rentier bargain', potentially resulting in social and political instability in the region. A recent IMF (2024, v) study on the fiscal stability of the Gulf region concluded that 'as global oil demand is expected to peak in the next two decades, the associated fiscal imperative could be both larger and more urgent than implied by the GCC countries' existing plans. At the current fiscal stance, the region's financial wealth could be depleted by 2034'. Luciani (2024, 406) sums up the situation thus 'The equilibrium of government budgets must be adapted to revenues that will be lower in absolute terms and differently composed, with oil rents playing a progressively reduced role.' Clearly, there are significant challenges ahead, but there are reasons to believe that the Gulf region will remain at the centre of the global oil and gas system for many years yet. But whose fossil fuels will be stranded by the high-carbon transition?

4.4 Whose fossil fuels get stranded?

This final section reviews the concept of the global carbon budget – the total emissions that can be released without pushing the global average temperature above a given threshold. Carbon Tracker Initiative (CTI, 2011) has played a significant role in promoting the concepts of the carbon budget, unburnable carbon, and stranded assets (see Box 4.2).

> **Box 4.2: The lexicon of unburnable carbon**
>
> **Carbon Budget**: To stabilize the global temperature at any level compared to pre-industrial, there is a finite amount of emissions that can be released before net emissions need to reach zero. For CO_2 emissions, this can be referred to as a carbon budget.
>
> **Unburnable Carbon**: Fossil fuel energy sources (reserves and/or resources), which physically cannot be burned if the world is to adhere to any given temperature outcome.
>
> **Stranded Assets:** Those assets that at some time before the end of their economic life (as assumed at the investment decision point), are no longer able to earn a financial return (that is, meet the company's internal rate of return), as a result of changes associated with the transition to a low-carbon economy (lower than anticipated demand/prices).
>
> Source: Carbon Tracker Initiative (2025, np)

Analysis of the remaining carbon budget and its relationship to fossil fuel combustion involves understanding the global resources and reserve base of each fossil fuel, along with its recovery costs and carbon intensity of production. The remaining carbon budget can then be linked to demand, price assumptions, and the supply cost curve to determine which resources and reserves must remain 'unburnt' to limit global temperature rise to a certain level over a defined period (Leaton, 2015). Linking back to the discussion in Chapter 2, normative scenarios of the global energy system transformation remain within the remaining carbon budget, measured in Gigatonnes of carbon (GtC). The concept of the remaining carbon budget is dynamic because, with every passing year, we use up an increasing portion of the remaining budget (Dalman, 2020). According to the Global Carbon Project (2024, nd) 'The remaining carbon budgets to limit global warming [with a 50% likelihood] to 1.5 °C, 1.7 °C and 2 °C are 65 GtC, 160 GtC, and 305 GtC respectively, equivalent to 6, 14 and 27 years from 2025.' Since 1850, 725 GtC has been emitted. As the UNEP

(2023, 2024a) makes abundantly clear in its *Production Gap* report (2023), the world is preparing to consume a level of fossil fuels that will rapidly deplete the remaining carbon budget.

4.4.1 The geography of unburnable fossil fuels

A body of scientific literature addresses the technical aspects of measuring and modelling the global carbon budget; however, for our purposes, it is sufficient to focus on research by colleagues at University College London (for a more comprehensive review, see Van der Ploeg and Rezai, 2020). The first paper by McGlade and Ekins (2015) was published shortly before the Paris Agreement was signed and presented an analysis of the geographical distribution of fossil fuels that must remain unused when limiting global warming to 2 °C. The timeframe for their analysis was 2010 to 2050. Their study concluded that 'globally, a third of oil reserves, half of gas reserves and over 80% of coal reserves should remain unused from 2010 to 2050 to meet the target of 2 °C' (McGlade and Ekins, 2015, 187). Although their paper was hugely influential, the Paris Agreement reset the goal of global climate action as limiting the rise in the global average temperature to less than 2 °C above preindustrial levels, and to as close to 1.5 °C as possible.

In 2021, Paul Ekins and colleagues at UCL published a follow-up paper calibrated to a 1.5 °C world (Welsby, Price, Pye, and Ekins, 2021). Logically, their analysis suggested that even more of the planetary resources and reserves of hydrocarbons would be 'unextractable' in a 1.5 °C world. They find that 'nearly 60% of oil and fossil methane gas and 90% of coal must remain unextracted to keep within the 1.5 °C carbon budget' (Welsby, Price, Pye and Ekins, 2021, 230). They defined unextractable fossil fuels as 'the volumes that need to stay in the ground, regardless of end use (that is, combusted or non-combusted), to keep within our 1.5 °C budget'. As with the earlier paper, they noted the 'disconnect between the production outlook of different countries and corporate entities and the necessary pathway to limit average temperature increases'. The analysis reveals a similar geography of unextractable carbon, with the Middle East, Russia, and other former Soviet states having the most significant influence on the global pattern. As with the earlier work, they determine that Arctic resources must remain undeveloped. The authors note the negative consequences of their findings for countries reliant on revenue from fossil fuel exports, and they also warn that their calculations are likely an underestimate, as they are based on a 50% chance of meeting the climate targets. Alongside the IEA's (2023) study on the role of the oil and gas industry in net-zero transitions, which informed COP-28, these prominent studies and the work of Carbon Tracker and others have sparked a significant debate about the transition risks linked to the high-carbon transition. However, the reality is that the phase-out of fossil fuels and the distribution of transition risks is unlikely to follow a straightforward, least-cost approach; instead, it will be a disorderly and heavily contested process that could lead to significant economic and political instability, especially among the fragile states mentioned earlier.

When it comes to the oil and gas industry, the majority of the analysis has focused on the strategies of the International Oil Companies (IOCs). However, according to the IEA (2023, 22), they own less than 13% of the oil and gas production and proved plus probable reserves. By comparison, the NOCs of the producer economies account for more than half of global oil and gas production and close to 60% of reserves. As noted earlier, the NOCs also hold the bulk of low-cost reserves, which should make them more resilient in the near term in the face of falling demand and prices; however, they all face significant long-term risks and will play a critical role in determining the outcome of the high-carbon transition (Manley and Heller, 2021; Furnaro and Yue, 2025).

4.4.2 Stranded assets and divestment

Those exposed to the transition risks associated with stranded assets include the reserve-holding states, companies involved in exploring, developing, and trading those reserves, as well as investors and shareholders who finance their activities. Given the significance of fossil fuel trade and investment to the global financial system, it is not surprising that the world's financial institutions are considering the implications of a large proportion of the world's oil and gas reserves being unextractable (Paun, Knight, and Chan, 2015). Again, a body of literature has developed on this issue, initially led by the CTI (Leaton, 2015) and researchers at Oxford's Smith School of Enterprise and the Environment (Ansar, Caldecott, and Tilbury, 2013). At the same time, these ideas have been marshalled by Environmental NGOs (ENGOs) to argue against new investments in fossil fuels (O'Connor, 2024). Not surprisingly, the oil and gas industry has questioned many of the claims made (IPIECA, 2014). Considering how the analysis of unburnable carbon functions, not all reserves are equal. Two factors are most important: first, the cost of extraction, and second, the carbon intensity of production. Both are complex issues involving geology, technology, and a wide range of above-ground factors, especially the extent of flaring and fugitive emissions. Kane (2018) observed that production costs are a strong indicator of carbon intensity, and Gulf producers benefit from low-cost production processes, flaring very little of the associated natural gas. Mehdi (2021, 3) suggested that this 'carbon efficiency' will be a key source of competitive advantage in the future.

In the past, when the oil and gas industry discussed 'advantaged reserves', it referred to low-cost, low-risk opportunities. Today, it has also come to signify low GHG intensity and a short life cycle, indicating they can be brought into production relatively quickly (Davies and Simmons, 2021). The IOCs aim to minimize both transition risk and the amount of GHG emissions on their carbon account, which financiers, regulators, and ENGOs increasingly scrutinise. It follows that if there are advantaged reserves, then there are also disadvantaged reserves, which are characterizd by high costs, long durations, high carbon emissions, and high geopolitical risks. Over the last decade, the IOCs have been disposing of assets that do not meet their criteria for profitability and sustainability (Arnold et al, 2023). Woodroffe (2024, 3) calculates that since 2014, approximately US$88 billion in assets have shifted from publicly listed companies to private – often local – companies, and the NOCs have acquired around US$24 billion in assets from non-NOCs. Furthermore, she observes that many of the acquired assets seem susceptible to transition risks. Two points are worth noting here: first, when the IOCs divest, those reserves do not stay in the ground; the new owners develop them. Second, for the producer economy – especially in emerging and developing markets – more of the transition risk is now borne by domestic players, who are unlikely to be as financially resilient as the IOCs.

4.5 Conclusions: Managing the fossil fuel incumbents

Historically, the literature on the resource curse has focused on how producer economies manage their heavy dependence on volatile rents from oil and gas resources. The very nature of the resource curse causes a series of problems that prevent many producer economies from effectively managing their rent dependence. Additionally, in the case of the Gulf States, a unique political economy has been developed to handle the distribution of rents, thereby supporting their autocratic regimes. Research on whose resources are likely to become stranded as the high-carbon transition accelerates indicates that there is potential for widespread economic and political instability among the most vulnerable states, which have also been the least equipped to handle the impact of the resource curse. There is a role here for national policy and international assistance to support those most affected by the shutdown of fossil fuel production. Without such support, a disorderly transition away from oil and gas could lead to a new round of conflict and instability.

FIVE

The Geopolitics of the Low-Carbon Transition

This chapter examines the geopolitical challenges associated with developing low-carbon energy systems. The acceleration of the low-carbon transition is crucial for mitigating the risks associated with reliance on fossil fuels (Bond, Butler-Sloss, and Walter, 2025). Rather than access to fuels, the low-carbon transition is about innovation and the production and deployment of a range of clean-energy technologies. The IEA (2024a) has identified seven key technologies: solar PV, wind, nuclear, electric vehicles, heat pumps, hydrogen, and CCS. Together, these technologies account for more than 75% of the emissions reductions in the IEA's APS and NZE scenarios. Here, I concentrate on electricity-based technologies rather than those that decarbonize fossil fuels, such as CCS. While these technologies may reduce emissions, they also perpetuate the geopolitical issues associated with fossil fuels. I also exclude nuclear power, which has its own unique geopolitical considerations. In this chapter, I examine a series of geopolitical challenges associated with what I earlier referred to as the 'build phase' of the low-carbon transition, which must be addressed to accelerate the transition.

The chapter is organized into three sections, each addressing what the IEA (2025a, 3) describes as 'emerging energy security risks': critical minerals, clean-energy supply chains, and electricity security. The first section, which makes up most of the chapter, focuses on the challenge of accessing critical materials that are vital inputs for clean technology. The second section examines the security of global supply chains for the technologies mentioned above, including a case study on batteries and electric vehicles (EVs). The third section introduces the concept of 'electricity security'. Electrification is central to energy system transformation, but increasing reliance on electricity introduces its own geopolitical risks.

5.1 The critical materials challenge

In its *2025 Global Critical Minerals Outlook*, the IEA (2025b, 10) states that 'In a world of high geopolitical tensions, critical minerals have emerged as a frontline issue in safeguarding global energy and economic security.' A considerable amount of research on this topic has been produced by international organizations (IEA, 2025b, 2024c; IRENA, 2023; World Bank, 2017, 2020), National Governments and their Geological Surveys (USGS, 2025; BGS, 2024), think tanks (Zhou and Månberger, 2024), investigative journalists (Sanderson, 2022), academics (Kara, 2024) and NGOs of various types (Global Witness, 2024). The nature of the challenge depends on your position in the supply chain. Resource-holding states aim to secure the best deal for the development of their mineral wealth; this means not only mineral rights but also capturing as much value-added for society as possible, while minimizing environmental and social impacts (see, for example, Government of Chile, 2023). They also seek to leverage the political influence that accompanies their increased importance. Importing, processing, and exporting nations, mainly China at present, aim to gain access to raw materials – from both domestic and international sources – to support their local mineral processing and consumption industries, as well as their export markets. Importing states have only recently realized that their clean tech industries, along with other high-tech sectors, such as electronics, IT, and the defence-industrial sector, are heavily reliant on supply chains that are mainly concentrated in a few states and are mostly controlled by China.

The issue of critical or strategic materials is not new. Wischer and Bazillian (2024) trace the importance of what they call 'mineral power' from the turn of the last century. At the height of the Cold War, the West was concerned about the impact of the Soviet natural resources on the global economy (Bradshaw and Connolly, 2016). Crebo-Rediker (2025) reports that at the end of the Cold War, the US Department of Defense sold off 99% of its national defence stockpile of critical materials and rare-earth elements. In the first decade of this century, a new wave of extractivism was driven by China's rapid industrial growth and its so-called 'going out' policy, whereby Chinese companies internationalized their supply chains to meet growing domestic demand (Veltmeyer and Petra, 2014). This process was then accelerated by the Belt and Road Initiative (Calderon et al, 2020, 7–8; Nedopil, 2025). The net result is the level of control that we see today with the West finding itself at least a decade behind China in developing the supply chains required to build its low-carbon energy systems, and much else.

In the analysis that follows, I focus on answering three questions: first, what are critical minerals and where are they mined and processed; second, how is the West responding to China's current dominance; and third, what are the consequences of this round of 'green' extractivism for the resource-holding states?

5.1.1 What are critical materials?

There is no generally accepted definition of 'critical minerals/materials', and there are a variety of terms that are used to denote a group of minerals and metals that have particular significance to industrial activity today: 'critical minerals', 'critical materials', 'critical raw materials', 'energy transition minerals', 'strategic minerals', and so on. The notion of 'criticality' is equally slippery. Critical for what, critical to whom, and critical for how long (Graedel, Gunn, and Espinoza, 2014). Riofrancos (2023, 24) observes that when it comes to lithium, for example, criticality is 'an emergent outcome of interacting variables: the discovery of deposits, the development of new extraction methods, government promotion of EVs, evolving battery chemistry, and recycling capacity'. At the same time, the critical materials challenge is framed in the context of heightened geopolitical competition and economic fragmentation, alongside an imperative to accelerate climate action (Kalantzakos, 2020). Current geopolitical tensions are also driving a cycle of rearmament in the West, and many energy transition materials are also critical to the defence-industrial sector (Hudson and Beaver, 2024; Vivoda, Matthews, and Andersen, 2025). The competition between decarbonization and militarization may amplify efforts to increase the supply of these materials.

For example, the Trump Administration's interest in critical materials is not driven by its green agenda; rather, it is the centrality of these materials to industrial competitiveness and defence that is the prime motivation. Table 5.1 maps a range of critical materials onto key low-carbon technologies. Some materials are critical to a narrow range of technologies, which the World Bank (2020, 12) refers to as 'concentrated materials'. The standout cases are lithium, graphite, and cobalt, which are critical to battery storage. Other materials are of more general significance, which the World Bank refers to as 'cross-cutting materials', the most obvious being copper. It is also the case that some technologies are more materials-intensive than others. Battery storage requires the greatest number of materials, comparable to those used in nuclear power. The table does not include the 17 elements of the periodic table called 'rare earths' (REEs) that are critical in the production of a wide range of clean technologies (Klinger, 2017; Kalaztzakos, 2020; Pitron, 2022). A final complication is that some of these materials are by-products of the production of other minerals.

THE GEOPOLITICS OF THE LOW-CARBON TRANSITION

Table 5.1: Mapping minerals with relevant low-carbon technologies

	Wind	Solar PV	Concentrated Solar	Hydro	Geothermal	Energy Storage	Nuclear	CCS	Total
Aluminium	X	X				X	X		4
Chromium	X			X	X	X	X	X	6
Cobalt						X		X	2
Copper	X	X	X	X	X	X	X	X	8
Graphite						X			1
Indium		X					X		2
Iron	X					X			2
Lead	X	X		X		X	X		5
Lithium						X			1
Manganese	X			X	X	X		X	5
Molybdenum	X	X		X	X		X	X	6
Neodymium	X								1
Nickel	X	X		X	X	X	X	X	7
Silver		X	X				X		3
Titanium				X	X		X		3
Vanadium						X	X		2
Zinc	X	X		X		X	X		5
Total	10	8	2	8	6	11	11	6	N/A

Source: Adapted from World Bank (2020, 37)

For example, indium – used in solar panels – is a by-product of zinc and lead mining, and cobalt is often a by-product of copper and nickel mining. This means that the demand for one mineral affects the availability of another.

Because certain materials are associated with particular technologies, there is significant variation in what is considered a critical material by different industries. Therefore, it is not surprising that different countries have varying lists of critical materials, reflecting their industrial structure. Research by the IRENA and NUPI (2024) analysed 35 existing lists from academic publications, reports from international and nongovernmental organizations and government documents. Only one category, REEs, appeared in all the lists. They then created a 'composite-index' that is presented in Table 5.2; the minerals are also ranked based on criticality (see the source

Table 5.2: NUPI/IRENA composite list of critical materials for the energy transition

Most critical for the energy transition in a global context (in order of criticality)
Lithium*, cobalt*, gallium*, rare earth elements (REEs)*, neodymium, indium, platinum group metals (PGMs), dysprosium, nickel, tellurium, praseodymium, graphite*, manganese*, copper and germanium*.
Moderately critical
Silver, strontium, platinum*, phosphorus, chromium, rhodium, lanthanum, ruthenium, aluminium*, boron/borate, selenium, palladium, cerium, vanadium, titanium* and silicon.
Least critical
Molybdenum, magnesium, yttrium, cadmium, terbium, zinc, iridium, zirconium, samarium, tungsten*, beryllium*, tin, iron/steel, europium, potassium, niobium, tantalum, gadolinium, lead and rhenium.

* On NATO's list of 12 defence-critical materials

Source: NUPI/IRENA (2024)

for their methodology). I have also indicated which materials are on NATO's list of 12 'defence-critical raw materials', to add yet another term. A final caveat: this discussion focuses on how geopolitics affects the availability of critical materials and, consequently, the pace of deploying low-carbon technologies (IRENA, 2023).

There is no shortage of resources in the Earth's crust; the problem is that turning these resources into reserves and then into production requires substantial capital investment and takes time, and it also comes with significant environmental and social impacts. An analysis of major mines that came online between 2010 and 2019 revealed that it took an average of over 16 years to develop projects from discovery to first production. However, the exact duration varies by mineral, location, and mine type (IEA, 2021, 129). One of the reasons for China's dominance has been its willingness to accept the environmental consequences of producing and processing these materials. A further complication is that the mining sector is also a contributor to the problem of climate change. In 2018, the GHG emissions from primary mineral and metals production were estimated to account for 10% of the total global energy-related GHG emissions (Azadi et al, 2020). Thus, there is a danger that a rapid expansion of mining activity to supply low-carbon technologies will contribute to increasing GHG emissions (IGF, 2024). Consequently, the international mining industry is under pressure to reduce its GHG emissions and broader environmental impacts, while also being called upon to increase investments and production to supply low-carbon technologies.

5.1.2 The geographical concentration of production and processing

As alluded to above, China's current control over critical minerals is perceived as a major geopolitical challenge by the West. Figure 5.1 illustrates the geographical concentration of

Figure 5.1: The geographical concentration of selected critical materials in 2023

[Bar chart showing composition by country for Copper, Lithium, Nickel, Cobalt, Natural graphite, and Rare earths. Legend: Rest of world, Brazil, United States, Myanmar, Mozambique, Russia, New Caledonia, Philippines, Indonesia, China, Peru, Democratic Republic of the Congo, Chile, Australia.]

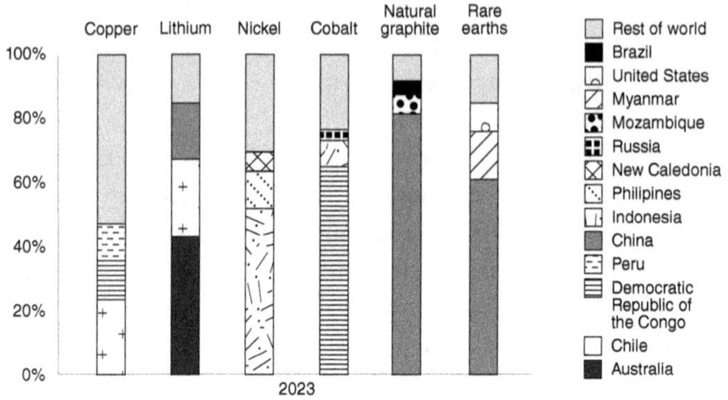

Note: Graphite extraction is for natural flake graphite. Rare earths are magnet rare earths only.
Source: IEA (2024c, 41)

mined or raw material production for selected critical materials in 2023 (the IEA did not update this figure in its 2025 report). The level of concentration is highest for natural graphite (China 64.6%) and REEs (China 70%), both of which are critical to battery storage (Klinger, 2017). IRENA (2023, 14) reports that two countries, Chile and Australia, account for 76.9% of lithium production, the Democratic Republic of the Congo (DRC) accounts for 70% of cobalt production, and Indonesia and the Philippines share 58.9% of nickel production. Only copper is relatively diversified, but even there, the top three producers – Chile, Peru, and DRC – account for 53.6% of production.

The geographical concentration of processing makes the level of Chinese dominance even clearer (Figure 5.2). Given that diversification is the golden rule of energy security, the high level of concentration within states is a key geopolitical risk, particularly if major reserve holders seek to use their dominant position to their advantage by limiting access, imposing duties, and/or manipulating markets, which China has on numerous occasions (IEA, 2025, 32). However, just as with fossil fuels, trade data give the false impression that it is countries that are trading with one another. The reality is that there is a complex ecosystem comprised of domestic mining companies, often state-owned, and international companies, both those involved in mining and processing and those that need these materials for manufacturing purposes. States clearly play a significant role, especially in granting access and controlling the trading regime, as well as being a source of investment and subsidies. International capital markets and the commodity exchanges that facilitate trade are also essential elements in the ecology of critical materials (Coe et al, 2025). However, it is not within their power to deliver critical materials themselves. The various minerals have different geographies and biographies; for example, when it comes to cobalt the DRC is centre stage, lithium has a different story to tell, with a locus in Latin America, while Indonesia and the Philippines are central to nickel and the tale of REEs is different again (Altiparmak et al, 2025; Chaudry, 2025; Cerai, 2024; Kalantzakos, 2017).

The prospect for particular materials is linked to specific technologies, and thus to the level of policy support required to produce and deploy those technologies. The current situation is characterized by price volatility (both high and low prices), often driven by inconsistent policy, coupled with significant medium- to long-term uncertainty. Two risks – technology and circularity – present significant challenges to the mining industry, particularly for private miners. Much like oil and gas, mining requires a substantial amount of investment, takes a long time and needs to pay off its investors. On the technology front, there is no guarantee that what is critical today will remain so in the future. For example, changing battery chemistry has reduced reliance on cobalt and more ubiquitous alternatives to lithium are now being pursued (e.g. sodium-ion batteries).

Figure 5.2: The geographical concentration of refined products, 2020 and 2024

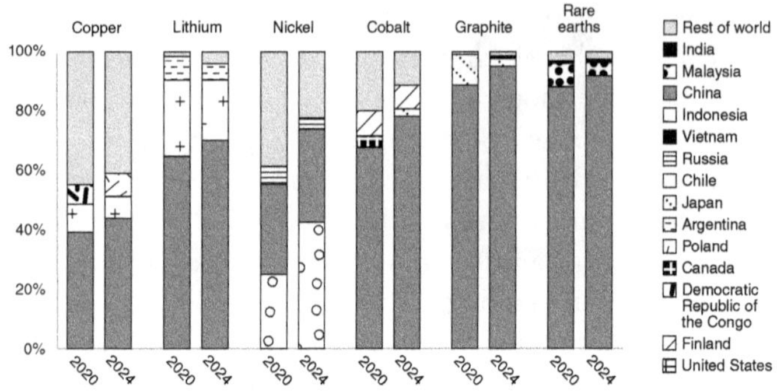

Note: Graphite is based on battery-grade spherical and synthetic graphite. Rare earths are magnet rare earths only.
Source: IEA (2025b, 27)

Notwithstanding this technology risk, the need to accelerate the build phase of the low-carbon energy system is creating a surge in demand that requires investment in new production. According to the IEA (2024a, 6–7), in their STEPS scenario, demand is projected to double between now and 2030, while it would triple in their APS scenario. Their NZE scenario would see demand quadruple from 10 million tons today to 40 million tons by 2040. Hammond and Brady (2022, 632) question whether near-term demand can be met when they observe that 'The bottom line is that brand new discoveries of critical mineral deposits leading to near-term production sufficient to meet 2030–2035 GHG objectives has very limited validity in practice.' Furthermore, a wave of new production could emerge just as recycling and circularity begin to reduce the need for mined production. However, it is unlikely that recycling alone will be sufficient, and not all materials can be recycled (Bloodworth, 2014). Walter et al (2024, 4) offer an alternative view, suggesting that 'Even as battery demand surges, the combined forces of efficiency, innovation and circularity will drive peak demand for mined minerals within a decade.' In the face of such uncertainty, it should be clear that meeting the critical materials challenge is not a simple case of 'Dig, baby, dig'.

5.1.3 How is the West responding to the challenge?

The story so far is that the western industrialized economies were happy to offshore the production of many of the materials that have now become critical to the low-carbon transition, in part because it was cheaper to do so, but also because their mining and processing is a dirty and energy-intensive process. Hudson and Beaver (2024) argue that the US is now incurring the cost of NIMBYism by not allowing domestic production of critical materials. China seized the opportunity to expand domestic production and develop global supply chains that integrate upstream mining operations at home and abroad with domestic processing and manufacturing industries, gaining the levels of dominance that we see today. A report by the UK Foreign Affairs Committee (2023, 3) aptly summarizes how the West perceives the challenge: 'Critical materials is not a geological challenge, but a geopolitical one. The vast majority of critical minerals are concentrated in countries that are autocratic, non-aligned, or actively hostile.' So, what are the geopolitical risks associated with the current situation?

Here, I have turned to IRENA's (2023) study on Critical Materials. It is worth noting that this study assesses the risks from the Western perspective. In the tradition of critical geopolitics, a narrative has emerged, as exemplified by the statement above, whereby China's current dominance is a threat to the interests of the West that must be countered; this is particularly so in the US. China has already shown a willingness to use its dominance of REEs for geopolitical ends on more than one occasion. The IRENA report (2023, 16) identifies six key geopolitical risks; it is worth noting that Chinese companies operating at home and abroad are just as exposed to some of these risks, although their behaviours are likely the cause of others.

- *External shocks*: Natural disasters, pandemics, wars, mine incidents, etc.
- *Resource nationalism*: Tax disputes, expropriation, foreign investment screening, etc.
- *Export restrictions*: Export quotas, export taxes, obligatory minimum export prices, licensing, etc.
- *Mineral cartels*: Co-ordination of production, pricing, market allocation, etc.
- *Political instability and social unrest*: Labour strikes, violence, corruption, etc.
- *Market manipulation*: Short squeezing, market cornering, spoofing, insider trading, etc.

In addition to domestic measures that include national assessments and strategies, the broader response of Western governments is often described as 'de-risking' and involves two processes: 'reshoring' and 'friendshoring' (Vivoda, 2023). In a broader context, globalization is based on the premise of comparative advantage; national economies should focus on production where they are globally competitive, and they should trade with other economies to import the things that they cannot produce competitively. Of course, the reality is far more complex, but the basic tenet is that free trade is more efficient and delivers greater economic prosperity. Liberals view the resulting interdependence as a positive-sum game in international relations. At the same time, realists perceive it as a zero-sum game, where one country's gain is at the expense of another (Kuzemko, Keating, and Goldthau, 2016, 8–12).

The West, and the US in particular, sees the current levels of geographical concentration and their resulting dependence as a geopolitical risk (Baskaran and Wood, 2025). For the US, this concern is part of an ongoing trade war with China. For more than a decade, there has been a consistent US policy of seeking to reduce dependence on imports from China of all kinds. Policies such as the Inflation Reduction Act, alongside Presidential Orders from both Trump and Biden, aim to develop domestic capacity to produce and process critical materials. This is reshoring. The EU has its European Critical Raw Materials Act, and national strategies are in place for countries such as Canada, India, Japan, Russia, and the UK. The EU approach is more balanced than is that of the US in terms of reshoring – aiming to expand production within the EU – while also establishing new trading relations, including with China, and promoting recycling and circularity (Kalantzakos, Overland, and Vakulchuk, 2023). It is essential to acknowledge that many non-Western states do not perceive the current situation as a threat and are willing to import low-cost Chinese technology, while some also welcome Chinese investment. However, by doing so, they are exposed to numerous risks, including those presented above (Chiengkul, 2025).

Friendshoring involves shifting sources of supply from countries seen as hostile to those regarded as allies or friends (Vivoda, 2023). This approach has been bolstered by the emergence of bilateral, plurilateral, and multilateral partnerships that aim to develop new sources of supply with lower geopolitical risks. An example of the bilateral approach is the US's deal with Ukraine to gain access to its mineral wealth. The US is also reportedly attempting to strike an agreement with the DRC in exchange for support in its conflict with rebel groups backed by its neighbour, Rwanda. IRENA (2023, 96) suggests that countries seeking secure access to critical materials can support international trade and investment agreements, promote regulatory cooperation (such as on standards), provide export credits for overseas mining investments, and engage in critical material diplomacy to create partnerships and even joint purchasing arrangements. IRENA (2023, 105) lists seven different international critical material alliances, most of which involve OECD member states, and only one of which includes producer states. However, countries such as Australia and Canada are significant producers of these resources. In response, the BRICS bloc is now developing its own critical minerals strategy (Müller, 2025).

The most significant alliance in terms of membership is the US-led Mineral Security Partnership (MSP) formed in 2022. It has 14 members. According to the US Department of State (2025), the MSP aims to accelerate 'the development of diverse critical minerals supply chains in cooperation with industry and other governments to support strategic projects and encourage investment throughout the value chain by reputable mining companies'. It is unclear whether the MSP will thrive under Donald Trump and his 'America First' approach. In addition to facilitating new production, these agreements also promote recycling, resource saving and substitute technologies; strategic stockpiling to mitigate short-term shortages. The targets of these alliances are the international mining companies that are expected to invest in new mining activities, as well as the reserve-holding states that are expected to make their resources available for exploration, development, and export. At face value, it sounds like the basis for a new round of unequal exchange and exploitation, so what are the consequences of green extractivism for the resource-holding states?

5.1.4 What are the consequences of green extractivism?

Some see the current round of green extractivism as a 'geoeconomic opportunity' for those emerging and developing economies that have significant critical material resources (IRENA, 2023, 92; Müller, 2023). The African Union's (AU, 2024) *African Green Mineral Strategy* exemplifies such optimism (Ouedraogo and Kilolo, 2024). The AU (2024, i) notes that 'Access to raw materials is a matter of strategic importance, raising the profile of mineral producing regions and increasing their bargaining power and opportunities for Equitable Resource Based Industrialisation.' However, it is well understood that mining has significant environmental impacts and is a potential source of social unrest and conflict (World Bank Group, 2004). Stewart (2025), at Global Witness, warns that 'Instead of enabling fair and just transitions for communities and countries globally, the race for critical materials is exacerbating human rights abuses, deepening inequality and fuelling global unrest.' These impacts amplify other problems, including food and water security, as well as forced migration. An estimated 54% of transition minerals are located on or near the land of indigenous peoples, many of whom, in theory, are protected by the UN Declaration on the Rights of Indigenous Peoples (Owen et al, 2022). The fear is that the current sense of urgency is promoting a new round of extractivism in the cause of the green transition that will lead to growing inequality and conflict as violent 'petrocultures' are replaced by 'green' energy and mineral security doctrines (Post and Le Billon, 2025; Debert and Le Billon, 2024). That said, Western governments are seeking to compete with China by offering an approach to extractivism that, they maintain, will promote positive environmental, social, and governance (ESG) outcomes. This role aligns with international mining companies that recognize the importance of ESG in securing a social license to operate. Despite their best intentions, international actors cannot mitigate the interests and actions of the governments of resource-holding states. While mining is an essential activity in many emerging and developing economies, the rents involved are modest compared to the sums generated by oil and gas; thus, there is unlikely to be a critical materials resource curse at the scale of the national economy of producer states (Hendrix, 2022). Unfortunately, there is already ample evidence to suggest that a new round of corruption and conflict is underway, sparked and fuelled by critical materials.

Church and Crawford (2018) have mapped the reserves of critical materials against measures of state fragility and concluded that 'a picture emerges of potential hotspots of increased fragility, conflict and violence resulting from growing mineral extraction'. They also presented a list of cases, many of which have gained notoriety since the study was conducted. For example, the case of cobalt in the DRC, where widespread corruption and violence are prevalent, and children are involved in artisanal mining (Amnesty International, 2023; Kara, 2024; Deberdt and Le Billon, 2024). In China, REEs production and the battery and solar supply chains are associated with widespread environmental hazards to public health and the reported use of forced labour involving ethnic minorities (Cranston et al, 2024). The ongoing conflict over lithium mining in Serbia and the concerns of the Sámi in relation to REEs in Sweden and Finland illustrate that this is not just a matter for emerging and developing states and that onshoring creates its own challenges (Larsen, 2023; Müller, Strack, and Vulovic, 2025). The EU's Critical Materials programme identifies 47 'strategic projects', most of which are located within the EU, that will enable Europe to meet its 2030 climate and digital objectives. What rights do local communities have in the face of such strategic projects? There is now a plethora of studies documenting the environmental and social wrongs being inflicted in the cause of the low-carbon transition (NRGI, 2025). Global Witness (2024) has been monitoring the incidence of violence and protests linked to critical materials mining, and it records that between 2021 and 2023, nearly 90% of violence and protests occurred in emerging economies, while companies from wealthier and major consuming economies conduct 81% of mining. Clearly, there are many parallels with the link between fossil fuels and conflict; so, what is to be done?

In April 2024, the UN Secretary-General launched the Panel on Critical Energy Transition Minerals, tasked with developing a set of common and voluntary principles to build trust, guide the transition, and accelerate the shift to renewable energy. In September 2024, the panel published its report (UN, 2024b) and presented seven principles, along with five actionable recommendations, all of which are laudable. Just after the Panel report was published in October, the UNEP (2024b) released a report on critical transitions, which included its own eight recommendations. In a similar vein, in February 2025, the IEA/OECD (2025) published a report on traceability in critical materials supply chains, with a pathway and eight steps. Clearly, there is no shortage of good intentions, and only time will tell if appropriate actions support them in addressing legitimate and growing concerns of the impacted communities about land rights and justice. Addressing these issues is essential for a just low-carbon transition.

5.2 Clean tech supply chains

Since the 1990s, the forces of economic globalization have promoted the creation of global value chains (GVCs) or global production networks (GPNs), aided by falling transport costs and support for free trade and foreign investment (Dicken, 2017). These GPNs and GVCs can be defined as 'organisational arrangements, coordinated by powerful lead firms, and linking suppliers, producers and states in the world economy' (Hess, 2021, 20). Much of the international trade between countries is internalized within these globally integrated production networks. A simple example is Apple's iPhone, which has famously carried the observation that it is designed in California and assembled in China. In the GPN literature, there is a robust discussion on the role of the state (Horner, 2016), which is primarily viewed as a facilitator of investment and production, but lately, there is increasing recognition of the impact of geopolitical risk in shaping state behaviour (Pavlinek, 2024; Solingen, 2025). Gong et al (2022, 168) suggest that GVCs and GPNs are being reconfigured by four forces: geopolitical uncertainties, climate change, technological change, and crises and shocks. Capri (2025, 11) evokes the GPN approach in his recent analysis of 'techno-nationalism', which he maintains describes how states seek to attain a competitive advantage for their stakeholders on both the local and global scales, and to leverage this advantage for geopolitical gain. This is not a new phenomenon; however, supply chains and networks are increasingly being influenced by state actions aimed at reducing national exposure to a range of geopolitical risks. When a state imposes tariffs and sanctions on other states, this impacts the integrity of related supply chains. Equally, efforts at reshoring and friendshoring change the locational choices open to those lead firms that are at the centre of global supply chains (Maihold, 2022). The US–China trade war during President Trump's first administration and the UK's vote for Brexit are seen as key moments of fragmentation. Then, the COVID-19 pandemic highlighted the fragility of globe-stretching supply chains, leading many companies to seek out more proximate suppliers where possible (near-shoring). However, as we shall see, that is currently problematic in the clean tech sector, where there are high levels of geographical concentration and dependence.

Figure 5.3 illustrates the IEA's (2024a) assessment of current levels of geographical concentration in some key technologies. Through its 'Made in China 2025' industrial strategy, introduced in 2015, China has sought to reduce reliance on foreign technology, enhance domestic innovation and build global competitiveness in strategic industries, which include clean tech (Boullenois, Black, and Rosen, 2025). The outcome is a high level of concentration in China, particularly in the solar PV sector. Each technology has its own story to tell. In the early days, Europe and the US had substantial solar PV industries. A combination of government support, a large domestic market, and over-capacity led to a rapid expansion of China's export capacity, with scale reducing costs (Carvalho, Dechezleprêtre, and Glachant, 2017). The initial response of the US and the EU was to try to protect domestic production through antidumping measures, but this effort ultimately failed. No doubt, the rest of the world sees the availability of cheap Chinese solar panels as a boon that is driving down the cost of the low-carbon transition and providing energy access. In other supply chains, such as wind energy, there is a more established industry in the US and the EU, and industrial strategies are

Figure 5.3: Geographical concentration of clean technology manufacturing capacity, 2023

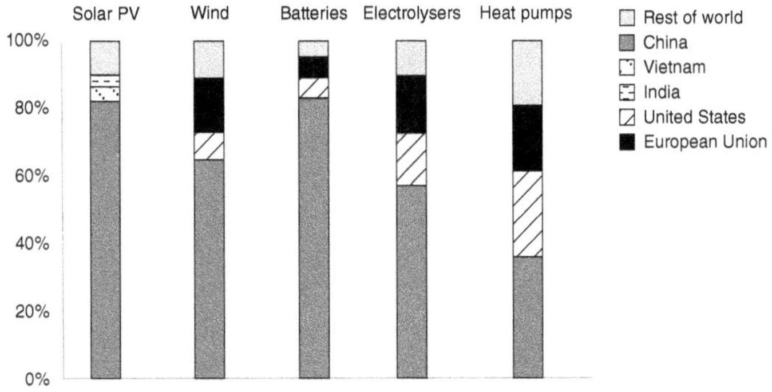

[Stacked bar chart showing Solar PV, Wind, Batteries, Electrolysers, Heat pumps with legend: Rest of world, China, Vietnam, India, United States, European Union]

Source: IEA (2025a, 31)

focused on nurturing and protecting these sectors (Van der Loos et al, 2022). Other sectors, such as electrolysers, are still in the early stages of development but also face competition from China (Ansari, Grinschgl, and Pepe, 2022).

These clean tech supply chains are also implicated in the growing geopolitical competition over semiconductor production, the 'Chip Wars', with significant production located in Asia, particularly in Taiwan (Miller, 2022; Wong et al, 2024). These technologies are also major consumers of critical materials and are essential components of clean tech. The US is particularly concerned about high-performance chips, implementing strict export controls and initiatives like the Chip Act to safeguard domestic production capacity. Asian producers – Japan, South Korea, and Taiwan – are also being encouraged to locate facilities in the US. This is a concern for Taiwan because its importance to the US mainly comes from its control of high-end chip manufacturing. If Taiwan were to lose that, would the US be as concerned with protecting it against Beijing's irredentism? Despite strict controls on the supply of the most powerful semiconductors to China, the country is demonstrating growing domestic competence in chip manufacturing and the associated development of artificial intelligence (AI) (Allen, 2023). Jaffe (2024, 179) summarizes the US position: 'As the new wave of digital technology innovation, driven by the convergence of automation, artificial intelligence, quantum computing, and big-data analytics remakes the world of green energy, it is a vital interest of the United States to maintain its leading role.'

Stepping back from the details, there is a discussion of 'deglobalization' and 'glocalization' as GPNs are reconfigured due to geopolitical constraints and geoeconomic competition (Rüger, Janssen, and Aulbur, 2021). Adopting an unashamed realist perspective, Capri (2025, 19) suggests that 'the world is not "deglobalizing"; instead, it's reorganizing and reconfiguring around opposing techno-nationalistic and geopolitical agendas'. This 'great bifurcation', he maintains, is dividing the world between the US and its allies, on the one hand, and China and its partners, on the other hand. This is a process sped up by the fallout from Russia's war in Ukraine but now made more complex by President Trump's trade war, which threatens to fracture the Western alliance. Batteries and EVs are ground zero in this 'Clean Tech Cold War' or 'Green War' (Post and Le Billon, 2025).

5.2.1 The case of batteries and EVs

Battery technology is essential to the low-carbon transition, both for reducing emissions in transportation and as a source of flexibility in future electricity systems, where it can eventually replace the reliable power supply currently provided by fossil fuels. To date, most demand has stemmed from the production of EVs, but this is likely to change as batteries are specifically designed for the power sector. This brief case study aims to show the complexity of GPNs and the challenges caused by geopolitics. The first two components of the EV battery supply chain were discussed above (Figure 5.4). The cell component stage involves the manufacture of specialized battery components. The cells

Figure 5.4: A simplified EV battery supply chain

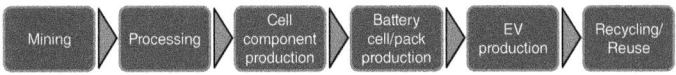

Source: Author

are then assembled to produce the battery pack, which also includes electronics, sensors, and the battery management system. These two stages are usually carried out within a gigafactory, which then supplies the original equipment manufacturer (OEM), better known as an automotive company. There are three critical points to note: first, this linear supply chain suggests the potential for a simple, integrated value chain; however, the reality is a complex ecosystem with many suppliers of materials and components at each stage. An entire ecosystem has developed around the production of cells and battery packs, with lead firms mainly based in China, Japan, and South Korea but operating globally. Consequently, there is now a complex GPN built around the production and supply of batteries for the EV market (more on this below).

Second, China's dominance is not by chance; instead, it results from a deliberate strategy by the Chinese Government to promote the development of what are called 'new energy vehicles' (NEVs) in China. The development of a world-leading NEV industry is a central pillar of China's desire to move up the value chain, away from being the workshop of the world, to leading the world in production and innovation in key high-value industrial sectors. The NEV sector had the benefit of low entry barriers compared to high-tech industries like semiconductor production (Capri, 2022, 12). True, China's automotive expansion began by inviting the major global OEMs to form joint ventures (JV) to produce in the domestic market. This is how Shanghai Motors (SAIC), most notably through its JV with Volkswagen, became the largest automotive manufacturer in China.

However, it is Chinese companies that dominate the global battery and EV market today, supported by government subsidies and a domestic market where vehicle ownership remains low by international standards, despite production now exceeding domestic demand. Strategically, the EV revolution offers multiple benefits to the Chinese state: firstly, it lowers oil demand (even though it increases electricity demand), enhancing energy security; secondly, it promotes investment, innovation, employment, and exports, which support economic growth; and finally, it improves air quality (although my recent experience in Shanghai suggests that it is also leading to more congestion). According to the IEA's (2025b, 16) most recent EV outlook, in 2024, almost half of China's car sales were electric, representing nearly two-thirds of the EVs sold globally. No matter how you measure it, China's battery and EV industry growth is impressive. Contemporary Amperex Technology Co. Ltd. (CATL) is the world's largest EV battery producer, and in 2025, Chinese EV manufacturers accounted for 70% of global production (IEA 2025, 12). In 2024, China's leading EV manufacturer, Build Your Dreams (BYD), surpassed Tesla (US) in sales, and in 2025, it will overtake Tesla in the number of cars built. BYD is also aiming for more than half of its sales to come from outside China in the near future. Third, this is an industry driven by technological change, focused on reducing costs and enhancing performance. Without going into details, battery chemistry is a clear example, with the Chinese automotive industry now favouring a different battery chemistry from Western OEMs and their battery supply chains. Every week, it seems like a company announces a new battery technology that will be game-changing. What is certain is that the battery technologies and EVs produced five or ten years from now will be very different from those on the road today. There may even be the possibility that some Western companies will leapfrog and regain a competitive advantage. The pace of change is a significant concern for governments, the industry, and consumers alike.

Bridge and Faigen (2022) have mapped the GPNs behind lithium-ion battery production. Their research makes clear just how complicated and intertwined the various stages in the production process are, and how firmly embedded Chinese and other Asian companies are in the activities of US and European OEMs. They describe battery production as 'an organisationally integrated, yet geographically dispersed process of materials production and assembly' (Bridge and Faigen 2022, 3). The gigafactory is at the heart of this process. Currently, there are approximately 240 gigafactories worldwide, and at the current growth rate, this number is expected to exceed 400 by 2030. Some of the states that supply critical materials, such as Chile, a key producer of lithium, would like to see gigafactories on their territory, as would Indonesia, a major producer of nickel. However, the OEMs prefer them to be located as close to vehicle manufacturing as possible. Host governments with established automotive industries are eager to offer incentives to secure their gigafactories. Already, Japanese and South Korean battery producers are well-established in North America and Europe, supplying the domestic automotive industry in these regions as well as in their home markets. Chinese producers are now forming joint ventures in Europe; for example, Stellantis and CATL are planning to build a gigafactory in Spain. Chinese OEMs are also establishing production facilities behind the EU's tariff barriers, with BYD operating in Hungary and planning further expansion. There is also Chinese ownership of European marques, with Volvo being the most high-profile example. This complicates the EU's intention to use tariffs to protect its domestic industry because it is part of a GPN that depends on China. A better option might be for Europe's OEMs to work with Chinese companies to expand their production in Europe.

As Bridge and Faigen (2022, 7) detail, there is a complex web of integration between companies across national borders. They note that 'The geopolitical consequences of expanding battery production extend beyond the security of mineral supply to the rapid deployment of gigafactories and the advancing electrification of the energy and mobility infrastructures to meet decarbonisation.' Capri (2022, 345) observes that 'The EV landscape is in complex flux, but one thing remains constant: it will continue to regionalise and localise as governments in China and the US and the EU each pursue their own agendas.' Considering current complexities, efforts by national governments to pursue techno-nationalism could potentially cause more harm than good, as they might cut domestic OEMs off from the GPNs they rely on. Thus, the rollback of Federal support for EVs in the US may leave the domestic industry behind as the global automotive sector electrifies.

Stepping back from the details and closing this discussion, it is clear that the world is at a crucial turning point. Whether due to deliberate neglect by the West or the Chinese government's resolve (probably both), China has secured a dominant position in producing some of the key supply chains and technologies vital to the low-carbon shift. Competition between the US and China has led the US to view this dominance as a threat to its national interests, prompting a push to develop domestic capacity rather than relying on Chinese imports. The term 'derisking' ('decoupling' in GPN-speak) is gaining popularity to describe how Western governments are intervening to promote greater diversity, while also seeking new economic opportunities through reshoring and friendshoring (Coe et al, 2025). But this process will take time. Europe finds itself in an uncertain position; it has economic interests to safeguard but also a clear goal to accelerate the decarbonization of its economies, partly to enhance energy security and partly to meet ambitious climate targets. Europe, therefore, faces a dilemma: to build or buy, or in other words, can it find a way to work with China for mutual benefit? If it can, this is likely to upset Washington. The rest of the world doesn't have to choose;

it can benefit from affordable clean tech supplies, improving energy access and accelerating the low-carbon transition.

5.3 Electricity security

In later years, 2025 might be seen as the year when 'electricity security' was acknowledged as a key aspect of future energy policy. It is widely accepted that 'electrification' is the path towards a decarbonized energy system. However, there is significant uncertainty over the pace of electrification and the ability of national power generation systems to keep pace with demand growth. One positive is that a more electrified energy system is more efficient than the current fossil-fuelled systems. A future electrified energy system is also cleaner, assuming that the power generation mix is decarbonized. The IEA estimates that in STEPS, the least ambitious of its three scenarios, global electricity demand will double by 2050, from 2,600 TWh in 2023 to 5,200 TWh in 2050 (IEA, 2024a, 39).

Furthermore, 80% of this demand growth is expected to occur in emerging and developing markets, with China accounting for approximately 40% of the global growth. Overall, the share of electricity in final energy consumption is likely to double from 20% today to over 40% in 2050. Recent developments in Europe and North America, such as the EU's REPowerEU programme and the expansion of data centre demand relating to AI, will accelerate electricity demand growth. At the same time, AI may enable more efficient management of electricity systems, potentially accelerating the transition to a low-carbon economy. However, AI will also enhance the efficiency of fossil fuel exploration and production, as well as mining. In a special report on AI and energy, the IEA (2025c, 14) notes that in 2024, data centres accounted for around 1.5% of global electricity consumption, primarily in the US (45%), China (25%), and Europe (15%). They forecast that data centre demand could more than double by 2030, primarily due to the rise of AI. Remember that data centres and AI are also major consumers of critical materials. What are the implications of all this for energy geopolitics?

It is fair to say that the question of electricity security has been neglected. There is a growing body of academic literature that contributes to a broader discussion on the geopolitics of renewable energy, alongside emerging analyses of the role of geopolitics in electricity interconnection (Smiith Stegen, Kusznir, and Riederer, 2024). However, in the policy arena, it is the IEA, again, that has championed the issue of electricity security in what they call the 'Age of Electricity'. The IEA (2021, 5) defined electricity security as 'the electricity system's capability to ensure uninterrupted availability of electricity by withstanding and recovering from disturbances and contingencies'. Table 5.3 presents key terms and definitions relating to operational security, adequacy, and resilience. In its background paper for the April 2025 Summit on the future of energy security (IEA, 2025a, 23), the IEA noted that 'The electricity security landscape today faces a complex array of interconnected challenges beyond traditional security concerns. Current vulnerabilities stem from premature retirement of dispatchable generation without adequate replacements, severe grid infrastructure constraints hampering integration of new resources, and increased price volatility affecting vulnerable consumers disproportionately.' A recent Eurelectric (2025, 8) report broadens electricity security of supply to include 'firm capacity and services, flexible capacity and services, transmission and distribution grids, and also raw material supply chains, resilience to climate change, physical and cyberattacks, geopolitical risks, etc'. This wider framing links concerns about electricity security to the discussion of critical materials and clean tech supply chains above. In an early academic discussion of grid interconnection, Fischhendler, Herman, and Anderman (2016, 541) suggested that 'electricity is a technical issue dealt with by economists, engineers and physicists' and that it is surprising that the perception is that 'electricity is not a matter of (geo)politics'. That perception is shifting, but as Pepe (2024, 4) notes, electricity security remains a complex issue; therefore, the way forward is likely to involve cross-disciplinary research merging geopolitical analysis with technical expertise. That is a project for another day; here, I am most interested in the relationship between geopolitics and the speed of electrification.

Table 5.3: Key electricity security terms and definitions

Term	Definition
Operational security	The ability of the electricity system to retain a normal state or to return to a normal state after any type of event as soon as possible.
Adequacy	The ability of the electricity system to supply the aggregate electrical demand within an area at all times under normal operating conditions. The precise definition of what qualifies as normal conditions and understanding how the system copes with other situations is key in policy decisions.
Resilience	The ability of the system and its component parts to absorb, accommodate and recover from both short-term shocks and long-term changes. These shocks can go beyond conditions covered in standard adequacy assessments.

Source: IEA (2021c, 5)

5.3.1 The three 'i's of electricity security

To keep things relatively simple, I want to introduce the notion of the three 'i's of electricity security: intermittency, integrity, and interconnection. Intermittency is the natural starting point for this discussion. While there is a dominant narrative that the rapid deployment of renewable power generation – particularly wind and solar – will reduce dependence on fossil fuels and be more secure because it is 'homegrown', it is also intermittent. There is some predictability, with solar PV and tidal power, for example. However, as the share of intermittent generation rises, it becomes necessary to have reliable access to flexible 'firm' (dispatchable) backup power generation. Electricity itself cannot be easily stored; therefore, it must be converted first, for example, into chemical energy in batteries or kinetic energy in pumped hydro and then converted back when needed. A technological breakthrough in this area would be truly game-changing.

Meanwhile, in most developed economies, fossil fuel power generation provides the majority of flexibility, while hydroelectricity and nuclear power also play a role. The exact power generation mix differs from country to country, and – as we will see – may also depend on regional interconnection. The main point here is that building a renewable power system requires careful planning, which relates to the second 'i', integrity.

In power systems, 'integrity' refers to the quality and reliability of the power delivery network, from its source to the load, ensuring stable voltage and current levels. In other words, the power system can keep the lights on. This links to the operational security and adequacy elements in Table 5.3. Without getting into the technicalities, there are at least three challenges to integrity that threaten electricity security.

First, there is the question of adequacy, which refers to the electricity system's ability to supply the aggregate electrical demand within an area at all times under normal operating conditions (Table 5.3). The notion of adequacy is more complex in a power system that is adding significant renewable power generation. Is there sufficient investment being made in power distribution networks – the grid – to connect new generation to consumers? The IEA (2023b, 8) suggests that building grid capacity at a sufficient pace risks becoming the weak link of clean-energy transitions. Planning and regulation are often viewed as barriers to connection. However, the statement from the IEA (2025a, 23) above highlights the dangers of not having sufficient dispatchable power generation capacity to support your 'greening grid'. The UK is a case in point, having removed coal from the power generation mix in late 2024, natural gas power is now the key source of flexibility and, for the present at least, that leaves electricity consumers hostage to global gas markets. The May 2025 Iberian blackout highlights the complexity of the challenge and the scale of its consequences when things go wrong.

Second, there are increasing physical threats to the integrity of power systems. They have long been vulnerable to damage from extreme weather, but the frequency and intensity of these events are growing, and more infrastructure is at risk (IEA, 2020, 47–53). Periods of extreme temperatures – both high and low – as well as drought and flooding – are occurring more often. Such events not only increase electricity demand but also threaten the operation of thermal power stations that require water for cooling, while reducing water availability for hydroelectric generation. Furthermore, there are ongoing climate-related threats to crucial power infrastructure, including rising sea levels and their impacts on coastal facilities. These issues are not geopolitical in origin, but they still pose challenges to energy security.

Third, there is increasing evidence of wilful damage to national power systems by hostile states. George et al (2025) report that in the days before Russia invaded on 24 February 2022, Ukraine had disconnected from the Russian power system to test its own energy self-sufficiency. However, on the day of the invasion, Russia disconnected Ukraine from its grid; however, in a matter of days, Ukraine was permanently connected to European grids. Since then, Ukraine's power infrastructure has been repeatedly targeted, and the country's power engineers have paid a heavy price while working to keep the system operational. In 2022 alone, it is reported that 98 power engineers were killed in the line of duty (UACrisis, 2023). In a special report on Ukraine, the IEA noted that

'Over the course of 2022–23, about half of Ukraine's power generation capacity was either occupied by Russian forces, destroyed or damaged, and approximately half of the large network substations were damaged by missiles and drones' (IEA, 2024c, 6). Hobhouse (2025) reports that by September 2024, Ukraine had lost 80% of its thermal power generation capacity due to Russian attacks and hydropower was also significantly degraded. A key lesson is that a centralized power system reliant on large thermal power stations, such as the Zaporizhzhia nuclear power plant, and associated infrastructures, is highly vulnerable when conflict gets kinetic. In response, Ukraine is rebuilding a more decentralized, renewable power system that is also more resilient to attack and easier to repair (Jermalavičius et al, 2025).

Even before Russia's renewed invasion of Ukraine, the country had been subjected to a series of cyberattacks in 2014–22 that aimed at degrading its power system. Cybersecurity is a growing concern in relation to electricity as grids become more complex and society becomes more reliant on electricity for its energy services (IEA, 2020, 38–46). A new study by EASAC (2025, 39–51) examines how Europe's increasingly digitized energy sector can be protected against the growing threat of cyberattacks and considers how the management of the system influences the security of electricity supplies. There are clear lessons to be learnt from the tragic circumstances in Ukraine, but also in a broader European context in the Baltic Sea, where there is increasing evidence of hostile states (China and Russia) damaging subsea cables that provide essential interconnections, both for power and data (O'Riordan, 2025). This concern is now spreading to the North Sea, prompting action by the littoral states, as well as by NATO and the EU (Majcin, 2025). All of this suggests an increasing 'securitization' of electricity security as the integrity of critical power infrastructures has become a matter of national security.

The final 'i' is interconnection. It is well understood that connecting to neighbouring power grids can enhance resilience.

This becomes particularly important in a renewable power system, as interconnection can offer flexibility and resilience that should, at least in theory, benefit all participants. However, as already discussed in the case of Ukraine, interconnection can be a source of vulnerability. After 15 years of effort, on 9 February 2025, the three Baltic states were finally disconnected from the Russian and Belarussian grids and integrated into the European market via Poland. Given that interconnection involves cooperation between states, it is perhaps not surprising that the geopolitics of electricity interconnection is the most developed issue of electricity security, though it remains relatively modest. In one of the earliest and more significant contributions, Westphal, Pastukhova, and Pepe (2022, 5) declared that 'Today, the impact of electricity interconnection deserves the closest possible scrutiny.' This, they maintain, is because it involves the interplay of three factors – 'the electricity grid, space and geopolitical power'.

Most of the work on interconnection has focused on examining particular geographical contexts, with Europe and the EU being the most studied, but there is also increasing attention being paid to Asia, and China is seeking to build connectivity through the BRI (Cornell, 2020; Setyawati and Nadhila, 2025). Scholten et al (2020) speculated that it could be the subject of great power rivalry in a world of 'grid communities, the size of continental super grids where prosumer countries operate an integrated electricity network and balance between secure domestic production and cheap imports'. There are mixed views on whether increased interconnection presents a geopolitical risk (Fang et al, 2024). Reviewing the successes and failures to date, it would seem that trust is a prerequisite for grid interconnection. For example, Fischhendler, Herman, and Anderman (2016) reviewed the repeated, failed attempts to build interconnection between Israel and its Arab neighbours. Elsewhere, cooperation remains modest, although there is no shortage of ambitious schemes. China envisions an Asian super grid, but it remains on the drawing board as it has yet to build a national grid at home. There is also evidence that interconnection can fall foul of regional and national politics. For example, the UK's exit from the EU has reduced the efficiency of trading between itself and its European neighbours, though there are hopes that this can be reset. In Norway, the impact of outflows through its interconnectors with the EU led to surging domestic electricity prices, prompting government intervention and the suspension of new interconnection projects. Slovakia threatened to cut off electricity supplies to Ukraine in protest of the EU's plans to stop all Russian gas imports by the end of 2027. Despite these challenges, Europe has established a highly integrated continental electricity market, which now accounts for approximately 80% of global cross-border electricity trade (Mills, 2025). This suggests that the rest of the world is not interconnected. Still, it is recognized that the acceleration of the low-carbon transition will require significant investments in interconnection to build a more resilient system. In North America, tariff wars between the US and Canada threaten long-standing cross-border electricity trading, another casualty of economic nationalism. The US itself has a very complex mosaic of regional grids, and Texas remains isolated from its neighbours.

Pepe (2024, 5) suggests that 'From a geostrategic point of view, fully interconnected grids create a web of interdependence, fostering diplomatic ties and ideally mitigating geopolitical tensions through shared energy resources.' I am not so sure. As noted, it would seem that electricity integration is unlikely to be a force for closer cooperation between neighbouring states unless there is already a high degree of trust and there are clear mutual benefits. Smith Stengen (2023, 1075) makes the interesting observation that 'The countries entering interstate grids gain benefits but lock themselves into long-term relations as the options for exiting are either to become isolated, which risks grid instability, particularly with variable renewables, or join a new grid.' As the possibilities for interconnection are geographically constrained, finding alternatives may not be that straightforward. Smith Stengen (2023, 1076) suggests that the density of integration is crucial to the durability of trading relations, with a degree of mutually assured destruction (MAD) required to maintain grid communities. In a world of increasing 'green electrification', interconnection can enhance efficiency, provide flexibility, and promote energy security. The problem is that, given the current levels of geopolitical fragmentation, mistrust and competition, it may prove challenging to increase the reach and density of grid connections just when they are needed the most. Even so, there is much that individual states can do to improve their electricity security. As the IEA (2025, 24) notes 'The most significant long-term risk [to electricity security], however, may be the missed opportunity to embed security and resilience considerations into today's system design decisions that will shape electricity infrastructure for decades to come.'

5.4 Conclusions: new systems, new challenges?

At the heart of the issues discussed above lie some fairly 'old school' geopolitical challenges. The high levels of geographical concentration in the production and processing of critical materials and clean tech have strengthened the states that have gained dominance and challenged those that now find themselves dependent. The problem is that there is no quick route to diversification. If states wish to accelerate the low-carbon transition, they will need to find ways of 'de-risking' the current situation while building alternative supply chains. This will require managing their current dependencies and implementing state intervention and support strategies, such as reshoring and friendshoring. However, this will be costly and time-consuming, contributing to a 'messy transition in the 2030s'. Far from delivering a peace dividend, there is a very real prospect that the build phase of the low-carbon transition will trigger new rounds of expropriation, conflict and inequality (Post and Le Billon, 2025).

When it comes to electricity security, the challenge of interconnection carries many of the same risks as 'pipeline geopolitics'. Countries are tied together through fixed, physical infrastructure – moving electrons instead of molecules – that is vulnerable to geopolitical leverage. Similarly, within states, decentralized renewable power generation and the development of local and regional grids will pose challenges to national cohesion. Furthermore, the technological path of the low-carbon transition remains uncertain, and new challenges and dependencies are likely to arise. It is only once the low-carbon system is established that its benefits, such as reduced geopolitical tensions and improved energy security, can potentially be realized. However, even then, electrification will bring its own set of geopolitical challenges to energy security.

SIX

Managing the Messy Mix

This guide explores the relationship between geopolitics and energy system transformation (EST). Specifically, it focuses on how geopolitical challenges affect the pace of transformation. The notion of a Messy Mix captures the complexity of the current moment, where we face all the challenges of the incumbent high-carbon system, made more complex by the prospect of falling demand, alongside an emergent set of challenges associated with the 'build phase' of the low-carbon transition. The pace of change is the critical uncertainty; it is also recognized that different parts of the world are at different starting points and have other priorities. Thus, the process of transformation varies geographically, and there is no single approach that fits all. The current pace of change does not suggest a gradual transition that is an extension of business-as-usual, nor is it an accelerated transition, as the rate of change is far off that required to achieve the goals of the Paris Agreement. Instead, the current geopolitical landscape suggests that the world is in the early stages of a messy transition that is likely to gain momentum and become more complex in the 2030s. Referring back to Figures 2.3 and 2.4 in Chapter 2, the Messy Mix does not come with its own set of challenges; rather, as Horizon 2, it is a period when both the challenges of phasing out fossil fuels (Horizon 1) and building out the low-carbon system are present (Horizon 3). In the words of Curry and Hodgson (2008, 3–2), Horizon 2 is 'an intermediate space in which the first and third horizons collide'. Over time, the challenges of the high-carbon transition will wane as the role

Table 6.1: The critical geopolitical challenges of energy system transformation

High-carbon transitions	Low-carbon transitions
• Managing the disorderly decline of demand for and supply of fossil fuels. • Ensuring a just and equitable transition for fragile fossil fuel producer economies.	• Diversifying critical materials supply chains. • De-risking clean tech global production networks. • Ensuring 'electricity security'.

of fossil fuels declines. Similarly, the challenges associated with the 'build phase' of the low-carbon transition will recede, and a new set of challenges will emerge as the global energy system becomes dominated by low-carbon energy (Horizon 3).

The world is at a critical inflection point; the physical risks of climate change are all too evident. Yet, at the same time, geopolitical rivalry, geoeconomic competition, and fragmentation are undermining the international cooperation required to enable climate action. In this context, I believe it is naïve to expect the world to follow a normative path where EST progresses smoothly according to the idealized policy and price assumptions of models; instead, the focus should be on managing the geopolitical challenges of this messy transition to accelerate change and promote more equitable and just transitions (Pastukhova and Walker, 2024). Table 6.1 presents the critical geopolitical challenges identified by this guide, which are combined during the Messy Mix; each is considered in more detail in the first two sections below. A final section then considers the challenges of managing the Messy Mix.

6.1 Geopolitics and the high-carbon transition

Fossil fuels dominate the current system and continue to shape the nature of energy geopolitics; however, limited thought has been given to understanding what will happen as fossil fuel demand declines. This section focuses on two challenges: the nature of the phase-out and the implications for producer economies.

6.1.1 Managing fossil fuel phase-out

The history of fossil fuels is one of cyclicality and volatility, and there is no reason to think that this will change. In fact, a permanent reduction (destruction) of demand is likely to make things even more volatile. Furthermore, there is no global governance structure in place, nor is one likely to be established, to manage the phase-out (Newell, 2021). As discussed in Chapter 4, the critical issue is which fossil fuels will remain in the mix the longest. The assumption is that coal will be the first to go, unsurprisingly, a view supported by the oil and gas industry. This is already the case in Europe and North America, but coal is likely to remain a significant presence in Asian markets for some time yet. While there is an understandable focus on the timing of peak fossil fuel demand and peak oil demand in particular – both are likely to be reached before 2030 – the key question is what happens thereafter? There is a significant difference between a peak and a plateau, and a peak that is followed by accelerating demand destruction. Fossil fuel incumbents and energy pragmatists are betting that demand will gradually fall in the 2030s and even increase for natural gas (some also think that oil demand will continue to grow). This will not only keep them in business but also legitimize investment in new production. While such an outcome would limit transition risk, the world would quickly exhaust the remaining carbon budget, thereby amplifying physical climate risk. It would also result in the continuation of fossil fuel geopolitics.

The alternative world is one of an accelerating phase-out of fossil fuels as low-carbon energy replaces fossil energy. How might this happen? Certainly not as a result of a global agreement to leave fossil fuels in the ground, however desirable that might be. Nevertheless, there is growing evidence that a range of supply-side approaches is gaining traction (Gaulin and Le Billon, 2020; Newell and Carter, 2024). Several 'climate clubs' have emerged, focusing on various sectoral aspects of implementing the Paris Agreement (Koppenborg, 2025). For example, the Powering Past Coal Alliance (PPCA) has set a target for its OECD members to phase-out unabated coal by 2030 and for its non-OECD members by 2040. The core members of the Beyond Oil and Gas Alliance (BOGA) aim to phase-out oil and gas production, and associate members seek to end subsidies and public funding for oil and gas exploration and production. Some advocates argue that 'no new fossil fuel projects' should become a global norm, asserting, correctly, that current reserves are more than sufficient to meet demand in a world of significant climate action (Green et al, 2024).

Rather than voluntary reductions in supply, the reality is that a transition away from fossil fuels will be driven by changes in demand, which are supported by government policies aimed at decarbonizing national economies and, in some cases, promoting energy security by reducing fossil fuel imports. The uptake of EVs is a critical indicator, and current developments in China may be a sign of what is to come. A study by the Rhodium Group (Quinn, 2025, np) has estimated that 'China's total electric vehicle fleet is already displacing over 1 million barrels per day in implied oil demand'. Thus, the rapid decarbonization of transport could soon have a significant impact on demand for crude oil and oil products. But electric vehicles are only as 'green' as the electricity they use. In some contexts, government mandates to decarbonize power generation will aid in powering past coal but may also maintain demand for natural gas. At the same time, it is also possible that some emerging and developing economies, including India, may eschew a gas transition, instead opting to leapfrog directly to renewable power generation. However, they may also continue to rely on coal for an extended period.

Cumulatively, these trends may result in a faster rate of fossil fuel demand destruction than envisioned by incumbents in the fossil fuel economy, potentially stranding assets and permanently reducing the rents earned from exports. But it will still fall short of the rate of change needed to mitigate physical climate risks. The consequences of this messy transition are uncertain. The literature on the geopolitics of EST suggests that the petrostates will lose their power and influence. At the same time, those essential to the low-carbon transition – so-called 'electrostates' – will become more significant (Mitrova and Corbeau, 2025).

6.1.2 A just and equitable high-carbon transition

Assuming that the world faces a disorderly transition away from fossil fuels in the 2030s, what are the consequences for the producer economies? Analysis in Chapters 3 and 4 made clear that not all petrostates will be challenged to the same extent. Much is dependent on the interplay between the scale of their reliance on fossil fuel rents and their global competitiveness in a world that favours low-cost and low-carbon intensity oil and gas production. Analysis suggests that there is a group of states that are highly dependent on fossil fuel rents, but unlikely to be competitive in a world of shrinking global demand. The historical volatility of international oil and gas markets suggests that it will not take much demand destruction to put markets into a tailspin as producers seek to carve out their share of a declining market.

The major producers with large, diversified economies will not be the first to feel the impact. A disorderly transition is expected to lead to increased conflict and inequality for the fragile fossil fuel producers – many of which are in MENA and Sub-Saharan Africa. These regions are also highly exposed to the physical impacts of climate change and are already facing significant economic and demographic pressures. In summary, not only is the high-carbon transition likely to be disorderly, but it is also likely to exacerbate existing injustices and inequalities, which have the potential to heighten geopolitical risks in already fragile regions. Much more needs to be done to understand and then mitigate the geopolitical risks associated with the fossil fuel phase-out (Gopalakrishan and Miller, 2024). Alongside encouraging producer economies to leave their fossil fuels in the ground, which they are unlikely to do, the emphasis should also be on supporting the most vulnerable states in managing their transition away from hydrocarbon exports. This is in the interest of all, as it is likely to result in a more just and less disorderly transition away from fossil fuels.

6.2 Geopolitics and the low-carbon transition

The dominant narrative on the geopolitics of renewables is that it will mean the end of fossil fuel geopolitics and improved energy security, as renewable power is homegrown. That may well be true of a future system. However, due to the nature of renewable energy systems, the primary challenges involve securing sufficient materials, manufacturing capacity, and funding to build and deploy the clean technologies necessary to accelerate decarbonization. As explained previously, this presents a series of familiar geopolitical challenges that stem from the geographical concentration of resources. In this case, the production and processing of critical materials. At the same time, the reliance on manufacturing and technology creates new challenges associated with clean tech supply chains. Furthermore, as the pace of electrification accelerates, a range of challenges must be managed to ensure energy security in the coming 'Age of Electricity'.

6.2.1 Diversifying critical materials supply chains

A significant amount of work has been done on mapping and analysing the current configuration of the production and processing of materials critical to the low-carbon transition. What is worrying for the West is that the latest IEA assessment of critical minerals shows that for some materials, the level of geographical concentration continues to increase (IEA, 2025b). So, what is to be done? Much depends on a state's location in the current geopolitical landscape. For those who possess significant reserves of critical materials, there is an opportunity to be seized, but it comes with numerous risks that must be managed to avoid a new round of extractive colonialism and unequal exchange.

For those dependent on imports, the harsh reality is that the current level of Chinese dominance cannot be quickly resolved. Sooner or later, all paths lead to China. This is perceived as a threat by the West because these materials are essential for their economic competitiveness, decarbonization efforts, and maintaining the military balance. As a result, a sense of urgency is prompting significant state intervention, alongside discussions of reshoring and friendshoring, accompanied by a rise in partnerships aimed at cooperation and encouraging investment. Nonetheless, as the previous chapter explained, considerable uncertainty remains about future demand for critical materials. Consequently, the international mining industry faces challenges similar to those of the oil and gas sector, requiring large-scale, long-term investments without guaranteed demand security, while also managing its greenhouse gas (GHG) emissions and environmental and social impacts. Simultaneously, evidence suggests that the current sense of urgency is fuelling a new wave of expropriation, exploitation, conflict, and inequality. Ignoring these challenges is entirely counterproductive, as it will not promote the necessary scale and pace of investment to diversify critical material supplies; instead, it will only complicate the low-carbon transition.

6.2.2 De-risking clean tech global production networks

High levels of geographical concentration are also prevalent in the manufacture of the clean tech critical to the low-carbon transition. Similar to manufacturers in Europe before them, those in China now benefit from a government-supported strategy that secures dominance through subsidies and the advantages of a large domestic market, which encourages scale economies and aids competitive exports. To date, the focus has been on technologies where entry costs are lowest; however, China is now moving up the value chain and gaining prominence in the next generation of clean technologies, including semiconductor production, data centres, and artificial intelligence (AI).

For many emerging and developing economies, gaining access to low-cost, clean technology is a welcome development, as long as it doesn't come with too many strings attached. However, for the West, it is seen as a serious threat to its geopolitical and geoeconomic competitiveness. The problem is that in some sectors, such as battery storage, China already has a significant lead that will be difficult to catch up to. Again, geopolitics matter. The US response to de-risking has been to engage in a trade war with China, protect domestic markets, and deploy state support (though some of this is now being curtailed) to build new domestic supply chains. However, this approach will take time and may not ensure competitiveness. Elsewhere, the West, broadly defined, faces a decision: if the emphasis is on accelerating the pace of the low-carbon transition – to address energy security and climate concerns – then it will have to find ways to de-risk current supply chains by cooperating with China. More broadly, it appears that the more geopolitical issues interfere with existing clean tech GPNs, the higher the cost of those essential technologies will be for the developed economies of the West. However, in a more divided world, China might do more to support the swift decarbonization of emerging and developing economies, along with improving their energy access. This suggests that it is the West that is experiencing the messiness of the low-carbon transition, while China is seizing the opportunity.

6.2.3 Ensuring electricity security

If critical materials have stolen the limelight in the geopolitics of the low-carbon transition, then electricity security is emerging stage left. For most developed economies, the strategy is to electrify as much as possible – starting with personal mobility and buildings – and then seek alternative solutions for those hard-to-abate sectors, including hydrogen and low-carbon drop-in fuels, as well as technologies to capture and sequester GHG emissions. For emerging and developing economies, electrification can provide a route to improved energy access without the dangers of carbon lock-in and the attendant fossil fuel insecurities, but the most significant challenge is securing the finance needed to fund electrification. It is noteworthy that today, China is the country with the fastest rate of electrification.

Electrification itself is inherently more efficient than traditional fossil fuel systems. But it should not be forgotten that there is also an imperative to improve efficiency and reduce demand, and thus, the size of the electricity system that needs to be built. In this context, much more can be done to promote efficiency and demand reduction as a means to improve energy security. At the same time, there are new sources of demand growth – most notably data centres and AI – that also need to be managed, though the latter can play a key role in enabling more efficient smart electrification. Furthermore, when developing new electricity infrastructure, adequate attention must be given to future resilience against both physical climate threats and geopolitical issues. Finally, there is the geopolitics of electricity interconnection. It remains to be seen whether interconnection can become a force for cooperation within and between states. But it is clear that in an age of electricity, a more fragmented and disconnected world will be less energy secure.

6.3 Managing the Messy Mix

The world is committed, through the UNFCCC/COP, to transitioning away from fossil fuels in a just, orderly, and equitable manner, to achieve net zero by 2050. But it is unclear how this might be achieved. The purpose of this guide is to explain the significant geopolitical challenges that could hinder efforts to accelerate the transformation of the energy system. The reality is that there is little chance of reaching an international agreement that will manage the phase-out of fossil fuels and speed up the development of the low-carbon energy system. As noted at the beginning of this guide, there is a significant right-wing populist backlash against decarbonization and net zero. However, responding by simply stating that the way to a more secure energy future is to phase-out fossil fuels and build a new system based on renewables is not enough. This is a complex and contested process that requires a more holistic approach to energy geopolitics and climate action. Rather than holding on to hope for an orderly transformation, states, market participants, the academy, civil society, and many others need to accept the inevitability of a messy transition and the challenges that are emerging as we enter this phase of the Messy Mix. For example, the plight of fragile fossil fuel producers is equally important as that of those who bear the adverse effects of mining critical materials. In both cases, a failure to address these challenges will lead to increased conflict and instability that hinder the growth in prosperity needed to finance climate action, slowing the pace of change. The essence of the Messy Mix, then, is that we need to simultaneously manage the problems of the decline of the old and the growth of the new to accelerate action.

References

Abdel-Fattah, P. (2023) *Rise of GCC Sovereign Wealth Funds: Magic Wand?* Al Bator Research Centre. 26 March. Available at: http://www.habtoorresearch.com/programmes/rise-of-gcc-sovereign-wealth-funds-magic-wand.

AU [African Union] (2024) *Africa's Green Minerals Strategy*. Addis Ababa: African Union. Available at: https://au.int/en/documents/20250318/africas-green-minerals-strategy-agms.

Ahrend, R. (2005) Can Russia Break the 'Resource Curse?' *Eurasian Geography and Economics*, 46(8): 584–609.

Aleklett, K. and Campbell, C.J. (2023) The Peak and Decline of World Oil and Gas Production. *Minerals & Energy*, 18(1): 5–20.

Allen, G.C. (2023) *China's New Strategy for Waging the Microchip Tech War*. Washington, DC: CSIS. Available at: http://www.csis.org/analysis/chinas-new-strategy-waging-microchip-tech-war.

Alsharif, N., Bhattacharyya, S., and Intartaglia, M. (2017) Economic Diversification in Resource Rich Countries: History, State of Knowledge and Research Agenda. *Resources Policy*, 52(C):154–64.

Alssadek, M. and Benhin, J. (2023) Natural Resource Curse: A literature Survey and Comparative Assessment of Regional Groupings of Oil-Rich Countries. *Resources Policy*, 84: 103741.

Altiparmak, S.O., Waters, K., Thies, C.G., and Shutters, T.S. (2025) Cornering the Market with Foreign Direct Investments: China's Cobalt Politics. *Renewable and Sustainable Energy Transitions*, 7: 100113.

Amnesty International (2023) *DRC: Power Change or Business as Usual? Forced Evictions at Industrial Cobalt and Copper Mines in the Democratic Republic of Congo*. London: Amnesty International. Available at: http://www.amnesty.org/en/documents/afr62/7009/2023/en/.

Ansar, A., Caldecott, B., and Tilbury, J. (2013) *Stranded Assets and the Fossil Fuel Divestment Campaign: What Does Divestment Mean for the Valuation of Fossil Fuel Assets?* Oxford: Smith School of Enterprise and the Environment, University of Oxford. Available at: http://www.smithschool.ox.ac.uk/sites/default/files/2022-03/SAP-divestment-report-final.pdf.

Ansari, D., Grinschgl, J., and Pepe, J.M. (2022) *Electrolysers for the Hydrogen Revolution*. Berlin: SWP Comment. Available at: http://www.swp-berlin.org/10.18449/2022C57/.

Aoun, M.-C. (2013) Oil and Gas Resources of the Middle East and North Africa: A Curse or a Blessing? In: Chevalier, J-M and Geoffron, P. (eds) *The New Energy Crisis: Climate, Economics and Geopolitics* (2nd edn). Basingstoke: Palgrave Macmillan.

Arnold, J., Lockman, M., Toledano, P., Brauch, M.D., Sen, S., and Burger, M. (2023) *Transferred Emissions are Still Emissions: Why Fossil Fuel Asset Sales Need Enhanced Transparency and Carbon Accounting*. New York: Columbia Center on Sustainable Investment & Sabin Center for Climate Change. Available at: https://scholarship.law.columbia.edu/sustainable_investment/14/.

Ashford, E. (2024) The Green Transition: Implications for Energy Security and Geopolitics. *Stimson*. 17 December. Available at: www.stimson.org/2024/the-green-transition-implications-for-energy-security-and-geopolitics/.

Ashford, E. (2022) *Oil the State and War: The Foreign Policies of Petrostates*. Washington, DC: Washington University Press.

Atkins, E. (2023) *A Just Energy Transition: Getting Decarbonisation Right in a Time of Crisis*. Bristol: Bristol University Press.

Auty, R.M. (1993) *Sustaining Development in Mineral Economies: The Resource Curse Thesis*. London: Routledge.

Auty, R.M. and Furlonge, H.I. (2019) *The Rent Curse: Natural Resources. Policy Choice and Economic Development*. Oxford: Oxford University Press.

Auty, R.M. and Gelb, A.H. (2001) Political Economy of Resource-Abundant States. In: Auty, R.M. (ed) *Resource Abundance and Economic Development*. Oxford: Oxford University Press.

REFERENCES

Azardi, M., Northey, S.A., Saleem, H.A., and Edraki, M. (2020) Transparency on Greenhouse Gas Emissions from Mining to Enable Climate Change Mitigation. *Nature Geoscience*, 13: 100–04.

Azzuni, A. and Breyer, C. (2018) Definitions and Dimensions of Energy Security: A Literature Review. *WIREs Energy and Environment,* 7(1): e268.

Bamidele, S. and Erameh, N.I. (2023) Environmental Degradation and Sustainable Peace in the Niger Delta Region of Nigeria. *Resources Policy*, 80: 103274.

Baskaran, G. and Wood, D. (eds). (2025) *Critical Minerals and the Future of the U.S. Economy*. Washington, DC: Centre for Strategic and International Studies. Available at: http://www.csis.org/analysis/critical-minerals-and-future-us-economy.

Beblawi, H. (1987) The Rentier State in the Arab World. In: Beblawi, H. and Luciani, G. (eds) *The Rentier State*. London: Routledge.

Beck, M. and Richter, T. (2021) Oil and the Political Economy in the Middle East: Overcoming Rentierism. In: Beck, M. and Richter, T. (eds) *Oil and the Political Economy in the Middle East: Post-2014 Adjustment Policies of the Arab Gulf and Beyond*. Manchester: Manchester University Press.

Bentham, J. (2014) The Scenario Approach to Possible Futures for Oil and Natural gas. *Energy Policy*, 64: 87–92.

BlackRock (2024) *Energy Pragmatism: An Evolving Approach for the Mid-21st Century*. New York-BlackRock. Available at: http://www.blackrock.com/corporate/literature/whitepaper/energy-pragmatism.pdf.

Blankenship, B., Lisko, C., Overland, I., Urpelainen, J., Vakulchuk, R., and Yang, J. (2024) When Do Petrostates Diversify Their Exports? Urgency, Interests and Policy Design in Egypt, Kazakhstan and Malaysia. *Development Policy Review*, 42(6): e12808.

Bloodworth, A. (2014) Track Flows to Manage Technology – Metal Supply. *Nature*, 505: 19–20.

BloombergNEF (2024) *New Energy Outlook 2024*. London: BloombergNEF. Available at: https://about.bnef.com/new-energy-outlook/.

Blondeel, M., Price, J., Bradshaw, M., Pye, S., Dodds, P., Kuzemko, C. et al (2024) Global Scenarios: A Geopolitical Reality Check. *Global Environmental Change*, 84: 102781.

Blondeel, M., Bradshaw, M.J., Bridge. G., and Kuzemko, C. (2021) The Geopolitics of Energy System Transformation: A Review. *Geography Compass*: e12580.

Bond, K., Bulter-Sloss, S., and Walter, D. (2025) *Energy Security in an Insecure World: The Electrostate Strategy*. London: Ember. Available at: https://ember-energy.org/app/uploads/2025/04/Slidepack-Energy-Security-in-an-Insecure-World.pdf.

Bond, K., Butler-Sloss, S., and Walter, D. (2024) *The Cleantech Revolution: It's exponential, Disruptive, and Now*. Basalt, CO: Rocky Mountain Institute. Available at: https://rmi.org/insight/the-cleantech-revolution/.

Boullenois, C., Black, M., and Rosen, D.H. (2025) *Was Made in 2025 China Successful?* New York: Rhodium Group. Available at: https://rhg.com/research/was-made-in-china-2025-successful/.

bp (2024) *Energy Outlook 2024*. London: bp. Available at: http://www.bp.com/en/global/corporate/energy-economics/energy-outlook/foreword.html.

Bradley, S., Lahn, G., and Pye, S. (2018) *Carbon Risk and Resilience: How Energy Transition is Changing the Prospects for Developing Countries with Fossil Fuels*. London: Chatham House. Research Paper. Available at: http://www.chathamhouse.org/2018/07/carbon-risk-and-resilience.

Bradshaw, M.J. and Boersma, T. (2020) *Natural Gas*. Cambridge: Polity Press.

Bradshaw, M.J. (2014) *Global Energy Dilemmas*. Cambridge: Polity Press.

Bradshaw, M. (2010) A New Energy Age in Pacific Russia: Lessons from the Sakhalin Oil and Gas Projects. *Eurasian Geography and Economics*, 51(3): 330–59.

Bradshaw, M. (2009a) The Geopolitics of Global Energy Security. *Geography Compass*, 3/5: 1920–37.

REFERENCES

Bradshaw, M. (2009b) The Kremlin, National Champions and the International Oil Companies: The Political Economy of the Russian Oil and Gas Industry. *Geopolitics of Energy*, 31(5): 4–14.

Bradshaw, M. and Connolly, R. (2016) Russia's Natural Resources in the World Economy: History, Review and Reassessment. *Eurasian Geography and Economics*, 57(6): 700–26.

Bradshaw, M.J. and Waterworth, A. (2018) Unconventional Trade-Offs? National Oil Companies: Foreign Investment and Oil and Gas Development in Argentina and Brazil. *Energy Policy*, 122: 7–16.

Breugel (2025) *European Natural Gas Imports*. Brussels: Breugel. Available at: https://www.bruegel.org/dataset/european-natural-gas-imports.

Bridge, G. and Faigen, E. (2022) Toward the Lithium-Ion Battery Production Network: Thinking Beyond Mineral Supply Chains. *Energy Research and Social Science*, 89: 102659.

Bridge, G. and Bradshaw, M. (2017) Making a Global Gas Market: Territoriality and Production Networks in Liquefied Natural Gas. *Economic Geography*, 93(3): 215–40.

Bridge, G. and Le Billon, P. (2017) *Oil* (2nd edn). Cambridge: Polity Press.

BGS [British Geological Survey] (2024) *UK Criticality Assessment*. Kenilworth: BGS. Available at: https://nora.nerc.ac.uk/id/eprint/538471/.

Bromley, S. (2006) Blood for Oil? *New Political Economy*, 11(3): 319–434.

Bros, T. (2012) *After the US Shale Gas Revolution*. Paris: Editions Technip.

Capri, A. (2025) *Techno Nationalism: How It's Reshaping Trade. Geopolitics and Society*. Hoboken, NJ: J Wiley & Sons.

Capri, A. (2022) *The Geopolitics of Electric Vehicles: Techno-Nationalism Reshapes the Automotive Industry*. Singapore: Hinrich foundation. Available at: http://www.hinrichfoundation.com/research/wp/tech/geopolitics-of-electric-vehicles/.

Carbon Tracker Initiative [CTI] (2025) 'Terms List'. London: CTI. Available at: https://carbontracker.org/resources/terms-list/.

CTI (2011) *Unburnable Carbon – Are the World's Financial Markets Carrying a Carbon Bubble?* London: CTI. Available at: https://carbontracker.org/reports/carbon-bubble/.

Calderon, J.L., Bazillian, M., Sovacool, B., Hund, K., Jowitt, S.M., Nguyen, T.P. et al (2020) Reviewing the Material and Metal Security of Low Carbon Energy Transitions. *Renewable and Sustainable Energy Reviews*, 124: 109789.

Carvalho, M., Dechezleprêtre, A., and Glachant, M. (2017) *Understanding the Dynamics of Global Value Chains for Solar Photovoltaic Technologies.* Economic Research Working Paper No. 40. World Intellectual Property Organization. Available at: http://www.wipo.int/publications/en/details.jsp?id=4231.

Cerai, A.P. (2024) Geography of Control: a Deep Dive Assessment on Criticality and Lithium Supply Chain. *Mineral Economics*, 37: 499–546.

Chaudary, M.S.A. (2025) Lithium Dreams. Local Struggles: Navigating the Geopolitics and Socio-Ecological Costs of a Low-Carbon Future. *Energy Research & Social Science*, 121: 103952.

Chiengkul, P. (2025) Towards a Greener BRI? Critical IPE, the Belt and Road Initiative, and Renewable Energy Transitions. *Australian Journal of International Affairs*, 79(3): 385–405, https://doi.org/10.1080/10357718.2025.2482719.

Chester, L. (2010) Conceptualising Energy Security and Making Explicit its Polysemic Nature. *Energy Policy*, 38: 887–95.

Cherp, A. and Jewell, J. (2014) The Concept of Energy Security: Beyond ohe Four As. *Energy Policy*, 75: 415–21.

Church, C. and Crawford, A. (2018) *Green Conflict Minerals: The Fuels of Conflict in the Transition to a Low-Carbon Economy.* London: International Institute for Sustainable Development. Available at: http://www.iisd.org/publications/report/green-conflict-minerals-fuels-conflict-transition-low-carbon-economy.

Clift, B. and Kuzemko, C. (2024) The Social Construction of Sustainable Futures: How Models and Scenarios Limit Climate Mitigation Possibilities. *New Political Economy*, 29(5): 755–69.

Climatewatch (2025) *Historical GHG Emission*. Climatewatch. Available at: http://www.climatewatchdata.org/ghg-emissions?breakBy=sector&end_year=2021&start_year=1990.

Clootens, N. and Ben Ali, M.S. (2021) The Resource Curse: How Can Oil Shape MENA Countries' Economic Development? In: Ben Ali, M.S. (ed) *Economic Development in the MENA Region: New Perspectives*. Cham: Springer Nature.

Coe, N.M., Sinclair, L., Gibson, C., and Warren, A. (2025) Resourcing GPNs: Multi-Scalar State Derisking of Energy Transition Minerals at the Time of a Polycrisis. *Journal of Economic Geography*, lbaf020.

Colgan, J.D. (2021) *Partial Hegemony: Oil Politics and International Order*. Oxford: Oxford University Press.

Colgan, J.D. (2013) Fuelling Fire: Pathways from Oil to War. *International Security*, 38(2): 147–80.

Collier, P. and Hoeffler, A. (2005) Resource Rents. Governance and Conflict. *Journal of Conflict Resolution,* 49(4): 625–33.

Conolly, R., Hanson, P., and Bradshaw, M. (2020) It's Déjà Vu All Over Again: COVID-19. The Global Energy Market and the Russian Economy. *Eurasian Geography and Economics*, 61(4–5): 511–31.

Corbeau, A.-S. (2025) *Bridging the US-EU Trade Gap with US LNG is More Complex Than it Sounds*. Center on Global Energy Policy. New York: Columbia University. Available at: http://www.energypolicy.columbia.edu/bridging-the-us-eu-trade-gap-with-us-lng-is-more-complex-than-it-sounds/.

Corbeau, A.-S., Downs, E., and Mitrova, T. (2025) *Power of Siberia 2: Russia's Pivot, China's Leverage, and Global Implications*. Center on Global Energy Policy. New York: Columbia University. Available at: http://www.energypolicy.columbia.edu/power-of-siberia-2-russias-pivot-chinas-leverage-and-global-gas-implications/.

Corden, W.M. (1984) Booming Sector and Dutch Disease Economics: Survey and Consolidation. *Oxford Economic Papers*, 36(3): 359–80.

Cornell, P. (2020) *International Grid Integration: Efficiencies, Vulnerabilities, and Strategic Implications in Asia*. Washington, D.C.: Global Energy Centre Atlantic Council. Available at: http://www.atlanticcouncil.org/wp-content/uploads/2020/01/Grid-Integration-final-web-version.pdf.

Cranston, C., Dorett, A., Martin, E., and Murphy, L.T. (2024) *Respecting Rights in Renewable Energy: Addressing Forced Labour of Uyghurs and Other Muslim and Turkic-Majority Peoples in the Production of Green Technology*. London: Modern Slavery and Human Rights Policy and Evidence Centre. Available at: http://www.antislavery.org/wp-content/uploads/2024/01/MSPEC_Uyghur_Research_Summary.pdf.

Crebo-Rediker, H. (2025) America's Most Dangerous Dependence: Washington Must Secure Supply of Critical Minerals that China Doesn't Control. *Foreign Affairs*. 7 May. Available at: http://www.foreignaffairs.com/united-states/americas-most-dangerous-dependence).

Curry, A. and Hodgson, A. (2008) Seeing in Multiple Horizons: Connecting Futures to Strategy. *Journal of Futures Studies*, 13(1): 1–20.

Dalman, A. (2020) *Carbon Budgets: Where Are we Now?* London: Carbon Tracker Initiative. Available at: https://carbontracker.org/carbon-budgets-where-are-we-now/.

Davidson, C. (2012) *After the Sheikhs: The Coming Collapse of the Gulf Monarchies*. London: Hurst.

Davies, A. and Simmons, M.D. (2021) Demand for 'Advantaged' Hydrocarbons During the 21st Century Energy Transition. *Energy Reports*, 7: 4483–97.

Deberdt, R. and Le Billon, P. (2024) Green Transition's Necropolitics: Inequalities. Climate Extractivism and Carbon Classes. *Antipode*, 56(4): 1264–88.

de Jong, M. (2023) Tracing the Downfall of the Nord Stream 2 Gas Pipeline. *WIREs Energy and Environment*, 13(1): e502.

de Jong, M. Van de Graff, T. and Haesebrouck, T. (2022) A Matter of Preference: Taking Sides on the Nordstream 2 Gas Pipeline Project. *Journal of Contemporary European Studies*, 30(2): 331–44.

REFERENCES

Dicken, P. (2017) *Global Shift: Mapping the Changing Contours of the Global Economy*, (7th edn). London: Sage.

Dittmer, J. and Sharp. S. (2014) General Introduction. In: Dittmer, J. and Sharp, J. (eds) *Geopolitics: An Introductory Reader*. London: Routledge.

Di Muzio, T. (2015) *Carbon Capitalism: Energy, Social Production and World Order*. London: Rowman and Littlefield.

Do, T.K. (2021) Resource Curse or Rentier Peace? The Impact of Natural Resource Rents on Military Expenditure. *Resources Policy*, 71: 101989.

Dodds, K. (2019) *Geopolitics: A Very Short Introduction*. Oxford: Oxford University Press.

Dodds, K. (2023) Why Venezuela is Threatening to Annex Guyana's Oil-Rich Province of Essequibo. *The Conversation*. 8 December. Available at: https://theconversation.com/why-venezuela-is-threatening-to-annex-guyanas-oil-rich-province-of-essequibo-219352.

Ekins, P. (2023) *Stopping Climate Change: Polices for Real Zero*. London: Routledge.

Emmerson, C. and Stevens, P. (2012) *Maritime Choke Points and the Global Energy System: Charting a Way Forward*. London: Chatham House. Available at: http://www.chathamhouse.org/sites/default/files/public/Research/Energy.%20Environment%20and%20Development/bp0112_emmerson_stevens.pdf.

EIA [Energy Information Administration] (2025) *A Look Back at Our Forecast for Global Crude Oil Prices in 2024*. Washington, DC: US EIA. Available at: http://www.eia.gov/todayinenergy/detail.php?id=64304.

EIA (2024a) *Country Analysis: World Oil Transit Chokepoints*. Washington, DC: US EIA. Available at: http://www.eia.gov/international/analysis/special-topics/World_Oil_Transit_Chokepoints.

EIA (2024b) *Oil and Petroleum Products Explained: Oil and Oil Exports*. Washington, DC: US EIA. Available at: http://www.eia.gov/energyexplained/oil-and-petroleum-products/imports-and-exports.php.

EIA (2023a) *Oil and Petroleum Products Explained: Where Our Oil Comes From?* Washington, DC: US EIA. Available at: http://www.eia.gov/energyexplained/oil-and-petroleum-products/where-our-oil-comes-from.php.

EIA (2023b) *What is OPEC+ and How is it Different from OPEC.* Washington, DC: US EIA. Available at: http://www.eia.gov/todayinenergy/detail.php?id=56420#.

EI [Energy Institute] (2025) *Statistical Review of World Energy 2024.* London: Energy Institute.

EI (2024) *Statistical Review of World Energy 2024.* London: Energy Institute.

E Intel. [Energy Intelligence] (2025) *Energy Transition Macro Outlook: Technology Drives Acceleration.* London: Energy Intelligence. Available at: http://www.energyintel.com.

EASAC [European Academies Science Advisory Council] (2025) *Security of Sustainable Energy Supplies.* Vienna: ESAC. Available at: https://easac.eu/publications/details/security-of-sustainable-energy-supplies-1

Equinor (2024) *2024 Energy Perspectives.* Stavanger: Equinor. Available at: http://www.equinor.com/sustainability/energy-perspectives.

Eurelectric (2025) *Redefining Energy Security in the Age of Electricity.* Eurelectric: Brussels. Available at: https://energy-security.eurelectric.org/wp-content/uploads/Eurelectric-report-energy-security-in-the-age-of-electricity-1.pdf.

European Commission (2025a) *REPowerEU: Affordable, Secure and Sustainable Energy for Europe.* Brussels: European Commission. Available at: https://commission.europa.eu/strategy-and-policy/priorities-2019-2024/european-green-deal/repowereu-affordable-secure-and-sustainable-energy-europe_en.

European Commission (2025b) *REPowerEU – 2 Years On.* Brussels: European Commission. Available at: https://energy.ec.europa.eu/topics/markets-and-consumers/actions-and-measures-energy-prices/repowereu-2-years_en.

REFERENCES

Eurostat (2025) *EU Trade with Russia – Latest Developments*. Brussels: Eurostat. Available at: https://ec.europa.eu/eurostat/statistics-explained/index.php?title=EU_trade_with_Russia_-_latest_developments.

Evans, M.J. (2017) Unconventional Hydrocarbons and the US Technology Revolution. In: Grafton, R.Q., Cronshaw, I.G. and Moore, M.C. (eds) *Risk. Rewards and Regulation of Unconventional Gas: A Global Perspective*. Cambridge: Cambridge University Press.

Fang, S. Jaffe, A.M., Loch-Temzelides, T. and Prete, L.P. (2024) Electricity Grids and Geopolitics: A Game-Theoretic Analysis of the Synchronization of the Baltic States' Electricity Networks with Continental Europe. *Energy Policy*, 188: 114068.

Fattouh, B. and Sen, A. (2021) Economic Diversification in Arab Oil Exporting Countries in the Context of Peak Oil and the Energy Transition. In: Luciani, G. and Moerenhout, T. (eds) *When Can Oil Economies Be Deemed Sustainable*. Cham: Springer Nature.

Fischhendler, I., Herman, L., and Anderman, J. (2016) The Geopolitics of Cross-Border Electricity Grids: The Israeli-Arab Case. *Energy Policy*, 98: 533–43.

Fishman, E. (2025) *Chokepoint: How the Global Economy Became a Weapon of War*. London: Elliott & Thompson Ltd.

Fouquet, R. (2024) The Digitalisation, Dematerialisation and Decarbonisation of the Global Economy in Historical Perspective: The Relationship Between Energy and Information Since 1850. *Environmental Research Letters*, 19: 014043.

Frankel, J. (2010) *The Natural Resource Curse: A Survey*. Discussion Paper 10–2. Cambridge, Mass: Harvard Environmental Economics Program.

Fressoz, J.-B. (2024) *More and More and More: An All-Consuming History of Energy*. London: Penguin Random House.

Furnaro, A. and Yue, G. (2025) The Geographies of National Oil Companies: From the Spatial Politics of Resource Control to Our Climate Futures. *Progress in Environmental Geography*, 4(3): 320–41.

Gaddy, C.G. and Ickes, B.W. (2005) Resource Rents and the Russian Economy. *Eurasian Geography and Economics,* 46(8): 559–83.

GECF [Gas Exporting Countries Forum] (2025) *Overview*. Doha: GECF. Available at: http://www.gecf.org/about/overview.aspx#.

Gaulin, N. and Le Billon, P. (2020) Climate Change and Fossil Fuel Production Cuts: Assessing Global Supply Side Constraints and Policy Implications. *Climate Policy*, 20(8): 888–901.

Gavin, D.A. and Levesque, L.C. (2005) *A Note on Scenario Planning*. Harvard Business School. 17 November, 9-306-003.

Geels, F. and Turnheim, B. (2022) *The Great Reconfiguration: A Social-Technical Analysis of Low-Carbon Transitions in UK Electricity, Heat, and Mobility Systems*. Cambridge: Cambridge University Press.

George, S., Yuan, M., Loutfi, F., Reyes, R., Tesfamichael, M., Dolan, A. et al (2025) Sharing Electricity across Borders Could Bolster Energy Security. *World Resources Institute Commentary*. 29 March. Available at: https://e-mc2.gr/el/news/sharing-electricity-across-borders-could-bolster-energy-security.

Global Witness (2024) *Briefing: Critical Mineral Mines Tied to 111 Violent Incidents and Protests on Average a Year*. London: Global Witness. Available at: https://globalwitness.org/en/campaigns/transition-minerals/critical-mineral-mines-tied-to-111-violent-incidents-and-protests-on-average-a-year/.

Goldthau, A. and Westphal, K. (2019) Why the Global Energy Transition Does Not Mean the End of the Petrostate. *Global Policy*, 10(2): 279–83.

Goldthau, A. and Sitter, N. (2015) *A Liberal Actor in a Realist World: The European Union Regulatory State and the Global Political Economy of Energy*. Oxford: Oxford University Press.

Gong, H., Hassink, R., Foster, C., Hess, M., and Garretsen, H. (2022) Globalisation in Reverse? Reconfiguring the Geographies of Value Chains and Production Networks. *Cambridge Journal of Regions, Economy and Society*, 15: 165–81.

Goodwin, P. and Wright, G. (2014) *Decision Analysis for Management Judgment* (5th edn). Hobokem, NJ: John Wiley & Sons.

Gopalakrishan, T. and Miller, J. (2024) New Climate Dis-Economies: The Political Economy of Energy Transitions in Fragile Fossil Fuel Producers. *Environment and Security*, 2(3): 348–74.

REFERENCES

Graedel, T.E., Gunn, G., and Espinoza, L.T. (2014) Metal Resources. Use and Criticality. In: Gunn, G. (ed) *Critical Materials Handbook*. Chichester: J. Wiley and Sons.

Gray, M. (2011) *A Theory of 'Late Rentierism' in the Arab States of the Gulf*. Center for International and Regional Studies. Georgetown University School of Foreign Service in Qatar. Occasional Paper No. 7. Available at: http://www.jstor.org/stable/10.1163/j.ctt1w8h356.10?seq=1.

Green, F., Bois von Kursk, O., Muttitt, G., and Pye, S. (2024) No New Fossil Fuel Projects: The Norm We Need. *Science*, 384(6699): 954–7.

Grigas, A. (2017) *The New Geopolitics of Natural Gas*. Cambridge. MA: Harvard University Press.

Gros, R., Hanna, R., Gambhir, A., Heptonstall, P., and Spiers, J. (2018) How Long Does Innovation and Commercialisation in the Energy Sector Take? *Energy Policy*, 123: 682–99.

Grübler, A. (2012) Energy Transitions Research: Insights and Cautionary Tales. *Energy Policy*, 50: 8–16.

Grübler, A. (2004) Transitions in Energy Use. *Encyclopaedia of Energy*, 6: 163–77.

Grübler, A., Wilson, C. and Nemet, G. (2016) Apples, Oranges and Consistent Comparisons of the Temporal Dynamics of Energy Transitions. *Energy Research & Social Science*, 22: 18–25.

Gustafson, T. (1989) *Crisis amid Plenty: The Politics of Soviet Energy under Brezhnev and Gorbachev*. Princeton: Princeton University Press.

Gustafson, T. (2012) *Wheel of Fortune: The Battle for Oil and Power in Russia*. Cambridge, Mass: Harvard University Press.

Gustafson, T. (2020) *The Bridge: Natural Gas in a Divided Europe*. Cambridge, Mass: Harvard University Press.

Gustafson, T. (2021) *Klimat: Russia in the Age of Climate Change*. Cambridge, Mass: Harvard University Press.

Hammond, D.R. and Brady, T.F. (2022) Critical Minerals for Green Energy Transition: A United States Perspective. *International Journal of Mining. Reclamation and Environment*, 36(9): 624–41.

Hanieh, A. (2018) *Money. Markest and Monarchies: The Gulf Cooperation Council and the Political Economy of the Contemporary Middle East*. Cambridge: Cambridge University Press.

Hanieh, A. (2024) *Crude Capitalism: Oil. Corporate Power and the Making of the World Market*. London: Verso Books.

Harris, F. (2024) The Geopolitics of LNG. In: Griffon, P. and Baker, D. (eds) *Liquefied Natural Gas: The Law and Business of LNG* (4th edn). Woking: Globe Law and Business. pp 115–30.

Helm, D. (2017) *Burn Out: The Endgame for Fossil Fuels*. New Haven: Yale University Press.

Hendrix, C.S. (2022) *Shift to Renewable Energy Could Be a Mixed Blessing for Minerals Exporters*. Policy Briefs. 22–1. Washington, DC: Peterson Institute for International Economics. Available at: http://www.piie.com/publications/policy-briefs/shift-renewable-energy-could-be-mixed-blessing-mineral-exporters.

Henderson. J. (2018) Russia's Gas Pivot to Asia: Another False Dawn or Ready for Lift Off? *Oxford Energy Insight: 40*. Oxford: OIES. Available at: http://www.oxfordenergy.org/wpcms/wp-content/uploads/2018/11/Russias-gas-pivot-to-Asia-Insight-40.pdf.

Henderson, J. (2024) The Impact of the Russian-Ukraine War on Global Gas Markets. *Current Sustainable/Renewable Energy Reports*, 11(1): 1–9.

Henderson, J. and Moe, A. (2019) *The Globalization of Russian Gas: Political and Commercial Catalysts*. Cheltenham: Edward Elgar.

Henderson, J. and Pirani, S. (eds). (2014) *The Russian Gas Matrix: How Markets are Driving Change*. Oxford: Oxford University Press.

Henderson, J. and Yermakov, V. (2024) Russian LNG: History and Prospects. In: Griffon, P. and Baker, D. (eds) *Liquefied Natural Gas: The Law and Business of LNG* (4th edn). Woking: Globe Law and Business.

Herb, M. (2014) *The Wages of Oil: Parliaments and Economic Development in Kuwait and the UAE*. Cornell: Cornell University Press.

Hertog, S. (2020) The 'Rentier Mentality'. 30 Years On: Evidence from Survey Data. *British Journal of Middle Eastern Studies*, 47(1): 6–23.

REFERENCES

Hertog, S. (2024) The Political Economy of Reforms under Vision 2030. In: Sfakianakis, J. (ed) *The Economy of Saudi Arabia in the 21st Century*. Oxford: Oxford University Press. 359–80.

Hertog, S. (2025) When Rentier Patronage Breaks Down: The Politics of Citizen Outsider on Gulf States' Labour Markets. *Studies in Comparative International Development*. https://doi.org/10.1007/s12116-024-09455-x.

Hess, M. (2021) Global Production Networks: The State, Power and Politics. In: F. Palpacuer F. and Smith, A. (eds) *Rethinking Value Chains: Tracking the Challenge of Global Capitalism*. Bristol: Bristol University Press.

Hobhouse, C. (2025) Keeping the Lights On: How Ukraine Can Build a Resilient Energy System (and Why This Matters to the EU. *European Institute for Security Studies*. 28 March. Available at: http://www.iss.europa.eu/publications/commentary/keeping-lights-how-ukraine-can-build-resilient-energy-system-and-why.

Högselius, P. (2012) *Red Gas: Russia and the Origins of European Energy Dependence*. Basingstoke: Palgrave Macmillan.

Högselius, P. (2023) The Political History of Fossil Fuels: Coal, Oil, and Natural Gas in Global Perspective. In: Scholten, D. (ed) *Handbook on the Geopolitics of the Energy Transition*. Cheltenham: Edward Elgar.

Hone, D. (2023) The End of Combustion? In: Sandaram, A.K. and Hansen. R.G. (eds) *Handbook of Business and Climate Change*. Cheltenham: Edward Elgar.

Horner, R. (2016) Beyond Facilitator? State Roles in Global Value Chains and Global Production Networks. *Geography Compass*, 11(2): e12307.

Huber, M.T. (2011) Oil, Life, and Fetishism of Geopolitics. *Capitalism Nature Socialism*, 22(3): 32–48.

Hudson, S. and Beaver, W. (2024) *Securing Critical Mineral Supply is a Defense Priority*. Washington, DC: Heritage Foundation. Available at: http://www.heritage.org/defense/report/securing-critical-mineral-supply-chains-defense-priority.

Humphreys, M., Sachs, J.D., and Stiglitz, J.E. (2007) *Escaping the Resource Curse*. New York: Columbia University Press.

IEA [International Energy Agency] (2018) *Outlook for Producer Economies 2018: What do Changing Energy Dynamics Mean for Oil and Gas Exporters?* Paris: OECD. Available at: http://www.iea.org/reports/outlook-for-producer-economies.

IEA (2020) *Power Systems in Transition: Challenges and Opportunities Ahead for Electricity Security*. Paris: IEA. Available at: www.iea.org/reports/power-systems-in-transition.

IEA (2021a) *Net Zero by 2050: A Roadmap for the Global Energy Sector*. Paris: IEA. Available at: http://www.iea.org/reports/outlook-for-producer-economies.

IEA (2021b) *The Role of Critical Minerals in Clean Energy Transitions*. Paris: IEA. Available at: http://www.iea.org/reports/the-role-of-critical-minerals-in-clean-energy-transitions.

IEA (2021c) *Analytical Frameworks for Electricity Security*. Paris: IEA. Available at: http://www.iea.org/reports/analytical-frameworks-for-electricity-security.

IEA (2022) *Security of Clean Energy Transitions*. Paris: IEA. Available at: http://www.iea.org/reports/security-of-clean-energy-transitions-2022.

IEA (2023) *Electricity Grids and Secure Energy Transitions: Enhancing the Foundations of Resilient, Sustainable and Affordable Power System*. IEA: Paris. Available at: http://www.iea.org/reports/electricity-grids-and-secure-energy-transitions.

IEA (2024a) *World Energy Outlook 2024*. Paris: IEA. Available at: http://www.iea.org/reports/world-energy-outlook-2024.

IEA (2024b) *Global Energy and Climate Model*. Paris: IEA. Available at: http://www.iea.org/reports/global-energy-and-climate-model.

IEA (2024c) *Global Critical Minerals Outlook 2024*. Paris: IEA. Available at: http://www.iea.org/reports/global-critical-minerals-outlook-2024.

IEA (2025a) *Summit on the Future of Energy Security: Background Paper*. Paris: IEA. Available at: https://iea.blob.core.windows.net/assets/445ad277-bb69-4053-933e-bb744de08ec7/BackgroundPaper_FutureofEnergySecurity_web.pdf.

REFERENCES

IEA (2025b) *Global Critical Materials Outlook 2025*. Paris: IEA. Available at: http://www.iea.org/reports/global-critical-minerals-outlook-2025.

IEA (2025c) *Global EV Outlook 2025*. Paris: IEA. Available at: http://www.iea.org/reports/global-ev-outlook-2025.

IEA/OECD (2025) *The Role of Traceability in Critical Minerals Supply Chains*. Paris: IEA/OECD. Available at: www.iea.org/reports/the-role-of-traceability-in-critical-mineral-supply-chains.

IEEFA [Institute for Energy Economics and Financial Analysis] (2025) *European LNG Tracker*. Valley City, OH: Institute for Energy Economics and Financial Analysis. Available at: https://ieefa.org/european-lng-tracker.

IEF [International Energy Forum] (2025) *IEF Outlooks Comparison Report*. Riyadh: IEF. Available at: http://www.ief.org/_resources/files/reports/ief-outlooks-comparison-report-2025.pdf?v=1.

IEP [Institute for Economics & Peace] (2024) *Global Peace Index 2024*. Sydney: IEP. Available at: http://www.visionofhumanity.org/wp-content/uploads/2024/06/GPI-2024-web.pdf.

IGF [Intergovernmental Forum on Mining. Minerals. Metals and Sustainable Development] (2024) *Decarbonization of the Mining Sector: Scoping Study on the Role of Mining in Nationally Determined Contribution*. Winnipeg: IISD. Available at: http://www.iisd.org/system/files/2024-08/igf-decarbonization-mining-sector.pdf.

IGU [International Gas Union] (2025) *2025 World LNG Report*. London: IGU. Available at: https://www.igu.org/igu-reports/2025-world-lng-report.

IPCC [Intergovernmental Panel on Climate Change] (2022a) The Evidence is Clear: The Time for Action is Now. We Can Halve Emissions by 2030. *IPCC Press Release*. 4 April. Available at: http://www.ipcc.ch/report/ar6/wg3/resources/press/press-release/.

IPCC (2022b) Summary for Policymakers. In: Shukla., Skea, J., Slade, R., Al Khourdajie, A., et al (eds) *Climate Change 2022: Mitigation of Climate Change. Contribution of Working Group III to the Sixth Assessment Report of the Intergovernmental Panel on Climate Change*. Cambridge: Cambridge University Press.

IPCC (2025a) *Inter-Governmental Panel on Climate Change*. Geneva: IPCC. Available at: http://www.ipcc.ch.

IPCC (2025b) *History of the IPCC*. Geneva: IPCC. Available at: http://www.ipcc.ch/about/history/.

IPIECA [International Petroleum Industry Environmental Conservation Association] (2014) *Exploring the Concept of 'Unburnable Carbon'*. London: IPIECA. Available at: http://www.ipieca.org/resources/exploring-the-concept-of-unburnable-carbon.

IRENA [International Renewable Energy Agency Authority] (2019) *A New World: The Geopolitics of the Energy Transformation*. Abu Dhabi: IRENA. Available at: http://www.irena.org/-/media/files/irena/agency/publication/2019/jan/global_commission_geopolitics_new_world_2019.pdf.

IRENA (2023) *Geopolitics of the Energy Transition: Critical Materials*. Abu Dhabi: IRENA. Available at: http://www.irena.org/Publications/2023/Jul/Geopolitics-of-the-Energy-Transition-Critical-Materials.

IRENA (2024) *Geopolitics of Energy Transition: Energy Security*. Abu Dhabi: IRENA. Available at: http://www.irena.org/Publications/2024/Apr/Geopolitics-of-the-energy-transition-Energy-security.

Jaffe, A.M. (2024) *Energy's Digital Future*. New York: Columbia University Press

Jermalavičius, T.J, Rõigas, H., Sukhodolia, O., and Teperik, D. (2025) *The Staying Power of Ukrainain Nights: Lessons of Wartime Resilience of the Electricity Sector*. Riga: International Centre for Defence and Security. Available at: https://icds.ee/en/the-staying-power-of-ukrainian-lights-lessons-of-wartime-resilience-of-the-electricity-sector/.

Jones, B. and Steven, D. (2015) *The Risk Pivot: Great Powers. International Security and the Energy Revolution*. Washington D.C.: Brookings Institution Press.

Jones Luong, P. and Weinthal, E. (2010) *Oil is Not a Curse: Ownership Structure and Institutions in Soviet Successor States*. Cambridge: Cambridge University Press.

REFERENCES

Kalantzakos, S. (2017) *China and the Geopolitics of Rare Earths.* Oxford: Oxford University Press.

Kalantzakos, S. (2020) The Race for Critical Materials in the Era of Geopolitical Realignment. *The International Spectator*, 55(3): 1–16.

Kalantzakos, S., Overland, I. and Vakulchuk, R. (2023) Decarbonisation and Critical Materials in the Context of Fraught Geopolitics: Europe's Distinctive Approach to a Net Zero Future. *The International Spectator*, 58(1): 3–23.

Kara, S. (2023) *Cobalt Red: How the Blood of the Congo Powers Our Lives.* New York: St Martin's Griffin.

Karl, T.L. (1997) *The Paradox of Plenty: Oil Booms and Petro-States.* Berkely: University of California Press.

Karl, T.L. (1999) The Perils of the Petrostate 'Reflections on the Paradox of Plenty'. *Journal of International Affairs*, 53(1): 31–48.

Kelanic, R.A. (2020) *Black Oil and Blackmail: Oil and Great Power Politics.* Ithica: Cornell University Press.

Kelly, P. (2016) *Classical Geopolitics: A New Analytical Model.* Stanford, CA: Stanford University Press.

Kemfert, C., Präger, F., Braunger, I., Hoffart, F.M., and Brauers. H. (2022) The Expansion of Natural Gas Infrastructure Puts Energy Transitions at Risk. *Nature Energy*, 7: 582–7.

Kivimaa, P. (2024) *Security in Sustainable Energy Transitions.* Cambridge: Cambridge University Press.

Klare, M. (2001) *Resource Wars: The New Landscape of Global Conflict.* New York: Henry Holt and Company.

Klare, M. (2004) *Blood and Oil.* New York: Henry Holt and Company.

Kleveman, L. (2003) *The New Great Game: Blood and Oil in Central Asia.* London: Atlantic Books.

Klinger, J.M. (2017) *Welcome to the Rare Earth Frontier.* Ithaca: Cornell University Press.

Krane, J. (2018) *Climate Strategy for Producer Countries: the case of Saudi Arabia.* Houston: Baker Institute for Public Policy, Rice University. Available at: http://www.bakerinstitute.org/research/climate-strategy-producer-countries.

Krane, J. (2019) *Energy Kingdoms: Oil and Political Survival in the Persian Gulf.* New York: Columbia University Press.

Kuzemko, C., Blondeel, M., Bradshaw, M., Bridge, G., and Fletcher. L. (2024) Rethinking Energy Geopolitics: Towards a Geopolitical Economy of Energy System Transformation. *Geopolitics*, 30(2): 531–65

Kuzemko, C., Keating, M.F. and Goldthau, A. (2016) *The Global Energy Challenge: Environment. Development and Security.* London: Palgrave.

Lahn, G. and Bradley, S. (2016) *Left Stranded? Extractives-Led Growth in a Carbon-Constrained World.* London: Chatham House, Research Paper. Available at: http://www.chathamhouse.org/sites/default/files/publications/research/2016-06-17-left-stranded-extractives-bradley-lahn-final.pdf.

Larsen, R.K. (2023) Materials Act Fails to Protect Sámi Rights – Here's How to Strengthen It. *Stockholm Environment Institute Perspective.* Stockholm: Stockholm Environment Institute. Available at: http://www.sei.org/perspectives/eus-critical-raw-materials-act-sami-rights-protection/.

Le Billon, P. (2001) *Fuelling War: Natural Resource Conflict and Armed Conflict.* London: International Institute for Strategic Studies, Adelphi Paper 373.

Le Billon, P. (2007) Geographies of War: Perspectives on 'Resource Wars'. *Geography Compass*, 1/2: 163–82.

Leaton, J. (2015) *The $ Trillion Stranded Assets Danger Zone: How Fossil Fuel Firms Risk Destroying Investor Returns.* London: Carbon Tracker Initiative. Available at: https://carbontracker.org/reports/stranded-assets-danger-zone/.

Lehman, T.C. (2017) The Geopolitics of Global Energy. In: Lehman, T.C. (ed). *The Geopolitics of Global Energy: The New Cost of Plenty.* Boulder, CO: Lynne Rienner Publishers Inc.

Losz, A., Boersma, T., and Mitrova, T. (2019) *A Changing Global Gas Order 3.0.* New York: Centre on Global Energy Policy. Columbia University. Available at: http://www.energypolicy.columbia.edu/publications/changing-global-gas-order-30/.

Luciani, G. (1987) Allocation vs. Production States: A Theoretical Framework. In: Beblawi, H. and Luciani, G. (eds) *The Reinter State.* London: Routledge.

REFERENCES

Luciani, G. (2024) Saudi Arabia – Beyond the Rentier State? In: Stakianakis, J. and Luciani, G. (eds) *The Economy of Saudi Arabia in the 21st Century*. Oxford: Oxford University Press.

Mahdavi, P. (2020) Institutions and the 'Resource Curse': Evidence from Cases of Oil-Related Bribery. *Comparative Political Studies*, 53(1): 3–39.

Mahdavy, H. (1970) The Patterns and Problems of Economic Development in Rentier States: The Case of Iran. In: Cook, M.A. (ed) *Studies in the Economic History of the Middle East*. Oxford: Oxford University Press.

Maihold, G. (2022) A New Geopolitics of Supply Chains: The Rise of Friend-Shoring. Berlin: *I Comment*. Available at: http://www.swp-berlin.org/publications/products/comments/2022C45_Geopolitics_Supply_Chains.pdf.

Majcin, J. (2025) Battle of the Baltic: Safeguarding Critical Undersea Infrastructure. *European Policy Centre Commentary*. 22 April. Available at: http://www.epc.eu/publication/Battle-of-the-Baltic-Safeguarding-critical-undersea-infrastructure-645780/.

Makarov, I., Chenh, H., and Paltsev, S. (2020) Impacts of Climate Change Policies Worldwide on the Russian Economy. *Climate Policy,* 10: 1242–56.

Malm, A. (2016) *Fossil Capital: The Rise of Steam Power and the Roots of Global Warming*. London: Verso Books.

Manley, D. and Heller, P.R.P. (2021) *Risky Bet: National Oil Companies in the Energy Transition*. London: Natural Resource Governance Institute. Available at: https://resourcegovernance.org/publications/risky-bet-national-oil-companies-energy-transition.

Marcel, V., Gordon, D., Ogeer, N., and Omonbude, E. (2023) Left Behind: Emerging Oil and Gas Producers in a Warming World. *Climate Policy*, 23(9): 1151–66.

McGerty, F. (2022) *Middle East Defence Spending: No Oil Bonanza?* International Institute for trategic Studies. 14 November. Available at: http://www.iiss.org/online-analysis/online-analysis/2022/11/middle-east-defence-spending-no-oil-bonanza/.

McGlade, C. and Ekins, P. (2015) The Geographical Distribution of Fossil Fuels is Unused When Limiting Global Warming to 2 °C. *Nature*, 517: 187–90.

McKibben, B. (2024) Roughly 40% of all Bulk Shipping Globally is for Coal, Oil and Gas. *Follow This*. 21 October. Available at: http://www.follow-this.org/roughly-40-percent-of-all-bulk-shipping-globally-is-for-coal-oil-and-gas.

McKinsey & Company (2024) *Global Energy Perspective 2024*. New York: McKinsey & Company. Available at: http://www.mckinsey.com/industries/energy-and-materials/our-insights/global-energy-perspective.

Medzhidova, D. (2022) Return of Coal: A Short Visit or a Long Stay? *BRICS Journal of Economics*, 3(4): 209–29.

Mehdi, A. (2021) The Middle East and the Geopolitics of the Energy Transition and Realities'. *Oxford Energy Forum*. Oxford: OIES. Available at: http://www.oxfordenergy.org/wpcms/wp-content/uploads/2021/02/THE-MIDDLE-EAST-AND-THE-GEOPOLITICS-OF-THE-ENERGY-TRANSITION-MYTHS-AND-REALITIES-.pdf.

Met Office (2025) *Effects of Climate Change*. Exeter: Met Office. Available at: http://www.metoffice.gov.uk/weather/climate-change/effects-of-climate-change.

Metreau, E., Young, K.E., and Eapen. S.G. (2024) World Bank County Classifications by Income Level for 2024–25. *World Bank Blog*. 1 July. Available at: https://blogs.worldbank.org/en/opendata/world-bank-country-classifications-by-income-level-for-2024-2025.

Mikellidou, C.V., Shakou, L.M., Boustras, G., and Dimopoulos, C. (2018) Energy Critical Infrastructures at Risk from Climate Change: A Start of the Art Review. *Safety Science*, 110(C): 10–20.

Miller, C. (2018) *Putinomics: Power and Money in Resurgent Russia*. Chapel Hill: University of North Carolina Press.

Miller, C. (2022) *Chip War: The Fight for the World's Most Critical Technology*. London: Simon & Schuster.

Mills, R. (2020) A Fine Balance: The Geopolitics of the Global Energy Transition in MENA. In: Hafner. M. and Tagliapietra, S. (eds) *The Geopolitics of the Global Energy Transition*. Cham: Springer Nature.

Mills, M.P., and Atkinson, N. (2025) *Energy Delusions: Peak Oil Forecasts: A Critique of Oil 'Scenarios' in the IEA World Energy Outlook 2024*. Washington, DC: National Center for Energy Analytics. Available at: https://energyanalytics.org/energy-delusions-peak-oil-forecasts/.

MOD [Ministry of Defence, UK] (2025) *Strategic Defence Review 2025*. London: MOD. Available at: http://www.gov.uk/government/publications/the-strategic-defence-review-2025-making-britain-safer-secure-at-home-strong-abroad.

Mitchell, T. (2011) *Carbon Democracy: Political Power in the Age of Oil*. London: Verso Books.

Mitrova, T. (2023) *The Outlook for Russia's Natural Gas Sector*. Russia's Global Energy Role. Fairfax VA: EIRP, Working Paper No. 5. Energy Innovation Reform Project. Available at: https://innovationreform.org/contact-us/.

Mitrova, T. and Corbeau, A.-S. (2025) *Petrostates and Electrostates in a World Divided by Fossil Fuels and Clean Energy*. The National Interest. 27 May. Available at: https://nationalinterest.org/blog/energy-world/petrostates-and-electrostates-in-a-world-divided-by-fossil-fuels-and-clean-energy.

Moritz, J. (2020) Rentier Political Economy in the Oil Monarchies. Kamrava, M. (ed) *Routledge Handbook of Persian Gulf Economies*. London: Routledge.

Müller, M. (2025) Between global geopolitics and national interests: BRICS cooperation in the mineral sector. *South African Journal of International Affairs*, 31 (1–2): 215–41.

Müller, M., Strack, L., and Vulovic, M. (2025) *The EU's Raw Materials Diplomacy: Serbia as a Test Case*. Berlin: SWP Comment No. 10. Available at: http://www.swp-berlin.org/en/publication/the-eus-raw-materials-diplomacy-serbia-as-a-test-case.

NASA [National Aeronautics and Space Administration. US] (2025) *Climate Change*. NASA. Available at: https://science.nasa.gov/climate-change/.

NATO [North Atlantic Treaty Organisation. Belgium] (2024) *NATO climate Change and Security Impact Assessment* (3rd edn). Brussels: NATO. Available at: http://www.nato.int/nato_static_fl2014/assets/pdf/2024/7/pdf/240709-Climate-Security-Impact.pdf.

Nedopil, C. (2025) *China Belt and Road Initiative (BRI). Investment Report 2024*. Brisbane: Griffith Asia ad Green Finance & Development Center. Available at: https://greenfdc.org/china-belt-and-road-initiative-bri-investment-report-2024/.

Newell, P. (2021) *Power Shift: The Global Political Economy of Energy Transitions*. Cambridge: Cambridge University Press.

NOAA [NOAA Global Monitoring Laboratory] (2025) *Trends in Atmospheric Carbon Dioxide (CO_2)*. Washington, DC: NOAA Global Monitoring Laboratory, US Department of Commerce. Available at: https://gml.noaa.gov/ccgg/trends/.

NRGI [Natural Resource Governance Institute] (2025) *Digging into the Problem: How Corruption Facilitates Socioenvironmental, Human Rights, and Indigenous Peoples' Rights Harms in the Mining Sector*. London: NRGI. Available at: https://resourcegovernance.org/publications/digging-into-the-problem.

NUPI [Norwegian Institute for International Affairs]/IRENA [International Renewable Energy Agency] (2024) *Constructing a Ranking of Critical Materials for the Global Energy Transition*. Oslo: NUPI. Available at: https://www.nupi.no/en/publications/cristin-pub/constructing-a-ranking-of-critical-materials-for-the-global-energy-transition.

Odell, P.R. (1986) *Oil and World Power* (8th edn). Harmondsworth: Penguin Books.

O'Connor, M. (2024) *Turning Tides: the Economic Risks of B.C.'s LNG Expansion in a Changing Energy Market*. London: Carbon Tracker Initiative. Available at: https://carbontracker.org/reports/turning-tides/.

OPEC [Organisation of Petroleum Exporting Countries] (2025) *Our Mission*. Vienna: OPEC. Available at: http://www.opec.org/opec_web/en/about_us/23.htm.

REFERENCES

O'Riordan, K. (2025) Geopolitics of Baltic Subsea Infrastructure. *Brussels Institute for Geopolitics*. 11 April. Available at: http://www.big-europe.eu/publications/2025-04-11-geopolitics-of-baltic-subsea-infrastructure.

O'Sullivan, M.L. (2017) *Windfall: How the New Energy Abundance Upends Global Politics and Strengthens America's Power*. New York: Simon & Schuster.

Ouedraogo, N.S. and Kilolo, J.M.M. (2024) Africa's Critical Minerals Can Power the Global Low-Carbon Transition. *Progress in Energy*, 6: 033004.

Our World in Data (2025) *Energy Data Explorer*. Oxford: Our World in Data. Available at: https://ourworldindata.org/explorers/energy.

Overland, A., Bazilian, M., Ilimbek Uuku, T., Vakulchuk, R., and Westphal, K. (2019) The GeGaLo Index: Geopolitical Gains and Losses After Energy Transition. *Energy Strategy Reviews*, 26: 100406.

Owens, J.R., Kemp, D., Lechner, A.M., Harris, J., Zhang, R., and Lebre, E. (2022) Energy Transition Minerals and Their Intersection with Land-Connected People. *Nature Sustainability*, 6: 203–11.

Palti-Guzman, L. and Eyl-Mazzega, M-A. (2023) *The Strategic Repositioning of LNG: Implications for Key Trade Routes and Choke Points*. Paris: Études De L'IFRI. Available at: http://www.ifri.org/en/studies/strategic-repositioning-lng-implications-key-trade-routes-and-choke-points.

Paltsev, S. (2016) The Complicated Geopolitics of Renewable Energy. *Bulletin of Atomic Scientists*, 72 (6): 390–5.

Pastukhova, M. and Walker, B. (2024) *An Orderly and Equitable Global Transition away from Fossil Fuels: An Action Framework to Navigate Economic, Financial and Geopolitical Volatility*. Berlin: E3G Briefing Paper. Available at: http://www.e3g.org/publications/an-orderly-and-equitable-global-transition-away-from-fossil-fuels/.

Paun, S., Knight, Z., and Chan, W-S. (2015) *Stranded Assets: What Next?* London: HSBC Global Research. Available at: http://www.longfinance.net/programmes/sustainable-futures/london-accord/reports/stranded-assets-what-next/.

Pavlinek, P. (2024) Geopolitical Decoupling in Global Production Networks. *Economic Geography*, 100(2): 138–69.

Pepe, J.M. (2024) *Europe and the Emerging Geopolitics of Electricity Grids*. Just Climate by FES. Bonn: Friedrich-Ebert-Stiftung e.V. Available at: https://library.fes.de/pdf-files/bueros/bruessel/21205.pdf.

Pitron, G. (2022) The Geopolitics of the Rare-Metals Race. *The Washington Quarterly*, 45(1): 135–50.

Post, E. and Le Billon, P. (2025) The 'Green War': Geopolitical Metabolism and Green Extractivism. *Geopolitics*, 30(2): 760–800.

Prince, G., Muttitt, G., Collett-White, R., and Coffin, M. (2023) *Petrostates of Decline: Oil and Gas Producers Face Growing Fiscal Risks as the Energy Transition Unfolds*. London: Carbon Tracker Initiative. Available at: https://carbontracker.org/reports/petrostates-of-decline.

Prokip, A. (2025) *Russia's Gas Transit through Ukraine: End of an Era? Focus Ukraine*. Kennan Institute. 4 February. Available at: https://gbv.wilsoncenter.org/blog-post/russias-gas-transit-through-ukraine-end-era.

Puyo, D.M., Panton, A., Sridhar, T., Stuermer, M., Ungerer, and Zhang, A.T. (2024) *Key Challenges Faced by Fossil Fuel Exporters during the Energy Transition*. Washington, DC: IMF, Staff Climate Note 2024/001. Available at: http://www.imf.org/en/Publications/staff-climate-notes/Issues/2024/03/26/Key-Challenges-Faced-by-Fossil-Fuel-Exporters-during-the-Energy-Transition-546066.

Pye, S., Bradshaw, M., Price, J., Zhang, D., Kuzemko, C.K., Sharples, J. et al (2025) The Global Implications of a Russian Gas Pivot to Asia. *Nature Communications*, 16: 386.

Quinn, R. (2025) Electric Trucks and the Future of Chinese Oil Demand. *Rhodium Group*. 1 July. Available at: https://rhg.com/research/electric-trucks-and-the-future-of-chinese-oil-demand/.

Richards, D. (2024) GCC: Lower Oil Prices Will Weigh on Budgets. *Emirates NDB*. 30 September. Available at: http://www.emiratesnbdresearch.com/en/articles/gcc-lower-oil-price-will-weigh-on-budgets.

Riofrancos, T. (2023) The Security-Sustainability Nexus: Lithium Onshoring in the Global North. *Global Environmental Politics*, 23(1): 20–41.

Ritchie, R. and Rosado, P. (2024) *Energy Mix*. Oxford: Our World in Data. Available at: https://ourworldindata.org/energy-mix.

Roberts, D.B. (2023) *Security Politics in the Gulf Monarchies: Continuity Amid Change*. New York: Columbia University Press.

Ross, M.L. (2001) Does Oil Hinder Democracy? *World Politics*, 53: 325–61.

Ross, M.L. (2004) What Do We Know About Natural Resources and Civil War? *Journal of Peace Research*, 41(3): 337–56.

Ross, M.L. (2012) *The Oil Curse: How Petroleum Wealth Shapes the Development of Nations*. Princeton: Princeton University Press.

Rosser, A. (2006) Escaping the Resource Curse. *New Political Economy*, 11(4): 557–70.

Rowell, A., Marriott, J. and Stockman, L. (2005) *The Next Gulf: London. Washington and Oil Conflict in Nigeria*. London: Constable & Robinson Ltd.

Roxburgh, C. (2009) The Use and Abuse of Scenarios. *McKinsey Quarterly*. 1 November. Available at: http://www.mckinsey.com/capabilities/strategy-and-corporate-finance/our-insights/the-use-and-abuse-of-scenarios.

Rüger, M.W., Janssen, R., and Aulbur, W. (2021) Rethinking Global Automotive Production Networks. *Roland Berger*. 25 March. Available at: http://www.rolandberger.com/en/Insights/Publications/Rethinking-Global-Automotive-Production-Networks.html.

Rutland, P. (2008) Russia as an Energy Superpower. *New Political Economy*, 13(2): 203–10.

Rutledge, I. (2005) *Addicted to Oil: America's Relentless Drive for Energy Security*. New York: I.B. Tauris.

Sachs, J.D. and Warner, A.M. (2001) The Curse of Natural Resources. *Natural Resources and Economic Development*, 45(4–6): 827–38.

Saha, D., Walls, G., Waskow, D., and Lazer, L. (2023) *Just Transitions in the Oil and Gas Sector: Considerations for Addressing Impacts on Workers and Communities in Middle-Income Countries*. Washington, DC: World Resources Institute. Working Paper. Available at: http://www.wri.org/research/just-transitions-oil-gas-sector-workers-communities-middle-income-countries.

Sakar, A. and de Waal, A. (2024) Going 'Cold Turkey': Oil Addiction and 'Traumatic' Decarbonization in Fragile Fossil Fuel Producers. *Environment and Security*, 2(3): 323–47.

Sanderson, H. (2022) *Volt Rush: The Winners and Loser in the Race to Go Green*. London: Oneworld.

Setyawati, D. and Nadhila, S. (2025) *Wired for Profit: Grid is the Key to Unlock ASEAN Energy Investment*. London: EMBER. Available at: https://ember-energy.org/latest-insights/wired-for-profit-grid-is-key-to-asean-energy-investment/.

Sharpe, B. (2020) *Three Horizons: The Patterning of Hope* (2nd edn). Axminster: Triarchy Press.

Sharpe, B., Hodgson, A., Leicester, G., Lyon, A., and Fazey, I. (2016) Three Horizons: A Pathway Practice for Transformation. *Ecology and Society*, 21(2): 47–62.

Scholten, D. (ed). (2018) *The Geopolitics of Renewables. Lecture Notes in Energy. Volume 61*. New York: SpringerNature.

Scholten, D. (2023) Introduction: The Geopolitics of the Energy Transition. In: Scholten, D. (ed) *Handbook on the Geopolitics of the Energy Transition*. Cheltenham: Edward Elgar.

Scholten, D. and Bosman, R. (2016) The Geopolitics of Renewables: Exploring the Political Implications of Renewable Energy Systems. *Technological Forecasting & Social Change*, 103: 273–83.

Scholten, D., Bazilian, M., Overland, I., and Westphal, K. (2020) The Geopolitics of Renewables: New Board. New Game. *Energy Policy*, 138: 111059.

Sharples, J.D. (2013) Russian Approaches to Energy Security and Climate Change: Russian Gas Exports to the EU. *Environmental Politics*, 22(4): 683–700.

REFERENCES

Sharples, J. (2024) *LNG Shipping Chokepoints: The Impact of Red Sea and Panama Canal Disruption*. Oxford: OIES, Paper: NG 188. Available at: http://www.oxfordenergy.org/wpcms/wp-content/uploads/2024/02/NG-188-LNG-Shipping-Chokepoints.pdf.

Sharples, J. (2025) *No Way Back? Challenges to Russian Pipeline Gas in Europe Make Near-Term Rebound Unlikely*. Oxford: OIES, Energy Insight: 166. Available at: http://www.oxfordenergy.org/publications/no-way-back-challenges-to-russian-pipeline-gas-in-europe-make-near-term-rebound-unlikely/.

Shell (2021) *The Energy Transformation Scenarios*. The Hague: Shell International B.V. Available at: http://www.shell.com/news-and-insights/scenarios/what-are-the-previous-shell-scenarios/.

Shell (2023) *The Energy Security Scenarios*. London: Shell Plc. Available at: http://www.shell.com/news-and-insights/scenarios/the-energy-security-scenarios.html.

Shell (2025) *The 2025 Energy Security Scenarios: Energy and artificial intelligence*. London: Shell Plc. Available at: http://www.shell.com/news-and-insights/scenarios/the-2025-energy-security-scenarios.html.

Skalamera, M. (2023) The Geopolitics of Energy after the Invasion of Ukraine. *The Washington Quarterly*, 46(1): 7–24.

Skea, J., van Diemen, R., Portugal-Pereira, J., and Al Khourdajie, A. (2021) Outlooks. Explorations and Normative Scenarios: Approaches to Global Energy Futures Compared. *Technological Forecasting & Social Change*, 168: 120736.

Smil, V. (2010) *Energy Transitions: History. Requirements*. Prospects. Denver, CO: Praeger.

Smil, V. (2016) Examining Energy Transitions: A Dozen Insights on Performance. *Energy Research & Social Science*, 22: 194–7.

Smil, V. (2017) *Energy and Civilisation: A History*. Cambridge, MA: MIT Press.

Smil, V. (2024) *Halfway Between Kyoto and 2050: Zero Carbon Is a Highly Unlikely Outcome*. Vancouver: The Fraser Institute. Available at: http://www.fraserinstitute.org/studies/halfway-between-kyoto-and-2050.

Smith Stengen, K. (2023) International Relations Theory on Grid Communities and International Politics in a Green World. *Nature Energy*, 8: 1073–7.

Smith Stengen, K., Kusznir, J., and Riederer, C. (2024) The Geopolitics of Energy Transportation and Carriers: From Fossil Fuels to Electricity and Hydrogen. In: Scholten, D. (ed) *Handbook on the Geopolitics of the Energy Transition*. Cheltenham: Edward Elgar.

Solingen, E. (2025) Global Value Chains in a Brave New World of Geopolitics. *Journal of Political Power*, 18(18): 112–14.

Sovacool, B.K. (2016) How Long Will it Take? Conceptualising the Temporal Dynamics of Energy Transitions. *Energy Research and Social Science*, 13: 202–15.

Sovacool, B.K. and Mukherjee, I. (2011) Conceptualizing and Measuring Energy Security: A Synthesised Approach. *Energy*, 36: 5343–55.

Sovacool, B.K., Baum, C., and Low, S. (2023) The Next Climate War? Statecraft. Security and Weaponization in the Low-Carbon Future. *Energy Strategy Reviews*, 45: 101031.

Sovacool, B.K., Geels, F.W., Dahl Anderson, D., Grubb, M., Jordan, A., Kern, F. et al (2025) The Acceleration of Low Carbon Transitions: Insights, Concepts, Challenges, and New Directions for Research. *Energy Research & Social Science*, 121: 1034948.

S&P Global Commodity Insights (2024) *The New Pragmatism: Scenarios to understand a volatile energy transition*. London: S&P Global Commodity Insights. Available at: http://www.spglobal.com/commodity-insights/en/news-research/special-reports/energy-transition/the-new-pragmatism-scenarios-to-understand-a-volatile-energy-transition.

Stanford, J. (2020) Mel Watkins and the Continuing Evolution of Staples Theory. *Studies in Political Economy*, 101(3): 280–87.

Statista (2024) Share of Extra-EU Natural Gas Import Value from Russia from 2010 to 2nd Quarter 2024. *Statista*. 24 November. Available at: http://www.statista.com/statistics/1021735/share-russian-gas-imports-eu/.

Stern, J. (1982) Spectres and Pipe Dreams. *Foreign Policy*, 48: 21–36.

REFERENCES

Stern, J. (ed) (2012) *The Pricing of Internationally Traded Gas*. Oxford: Oxford University Press.

Stern, N. (2014) *The Economics of Climate Change: The Stern Review*. Cambridge: Cambridge University Press.

Stevens, P., Lahn, G., and Kooroshy, J. (2015) *The Resource Curse Revisited*. London: Chatham House, Research Paper. Available at: http://www.chathamhouse.org/sites/default/files/publicati ons/research/20150804ResourceCurseRevisitedStevensLahnKo oroshyFinal.pdf.

Stewart, E.I. (2025) The Critical Minerals Scramble: How the Race for Resources is Fuelling Conflict and Inequality. *Global Witness*. 20 March. Available at: https://globalwitness.org/en/campaigns/transition-minerals/the-critical-minerals-scramble-how-the-race-for-resources-is-fuelling-conflict-and-inequality/.

Strojny, J., Krakiwiak-Bal, A., Kanaga, A., and Kocarzyk, P. (2023) Energy Security: A Conceptual Overview. *Energies*, 16: 5042.

Tan, D., Low, C.T., and Kustsunai Lam, S. (2024) *Crude Awakening: Fast rising seas threaten seaborne oil & energy security*. Hong Kong: China Water Risk. Available at: https://chinawaterrisk.org/wp-content/uploads/2024/05/CWR-2024-Crude-Awaken ing-Fast-rising-seas-threaten-seaborne-oil-and-energy-security-Spotlight-Japan-and-South-Korea-FINAL.pdf.

Terzi, A. and Fouquet, R. (2023) The Green Industrial Revolution: Lessons from the History of Past Energy Transitions. *EcoPol Forum*, 24(6): 16–22.

Thompson, E. (2022) *Disorder: Hard Times in the 21st Century*. Oxford: Oxford University Press.

TotalEnergies (2024) *TotalEnergies Energy Outlook 2024*. Paris: TotalEnergies. Available at: https://totalenergies.com/news/press-releases/totalenergies-energy-outlook-2024.

Trading Economics (2025) EU Natural Gas TTF. *Trading Economics*. 26 March. Available at: https://tradingeconomics.com/commod ity/eu-natural-gas

UACrisis (2023) 98 Power Engineers Died in Ukraine in 2022. *Ukraine Crisis Media Centre*. 21 February. Available at: https://uacrisis.org/en/za-2022-rik-zagynuly-98-energetykiv.

Umbach, F. (2017) Geopolitical Dimensions of Global Unconventional Gas Perspectives. In: Grafton, R.Q., Cronshaw, I.G. and Moore, M.C. (eds) *Risks, Rewards, and Regulation of Unconventional Gas: A Global Perspective*. Cambridge: Cambridge University Press.

UK Foreign Affairs Committee (2023) *A Rock and a Hard Place: Building Critical Mineral Resilience*. London: House of Commons. Available at: https://committees.parliament.uk/publications/42569/documents/211673/default/.

UN (2024) *Resourcing the Energy Transition: Principles to Guide Critical Energy Transition Minerals Towards Equity and Justice*. New York: UN. Available at: http://www.unep.org/resources/report/resourcing-energy-transition.

UNEP [United Nations Environment Programme] (2023) *Production Gap Report 2023: Phasing Up or Phasing Down?* New York: UNEP. Available at: http://www.unep.org/resources/production-gap-report-2023.

UNEP (2024a) *Emissions Gap Report 2024: No More Hot Air … Please!* Nairobi: UNEP. Available at: http://www.unep.org/resources/emissions-gap-report-2024.

UNEP (2024b) *Critical Transitions: Circularity, Equity, and Responsibility in the Quest for Energy Transition Minerals*. New York: UNEP. Available at: http://www.unep.org/resources/publication/critical-transitions-circularity-equity-and-responsibility-quest-energy.

UNFCCC [United Nations Framework Convention on Climate Change] (2023) *Conference of the Parties Serving as the Meeting of the Parties to the Paris Agreement. Fifth Session: First Global Stocktake*. New York: UNFCCC. Available at: https://unfccc.int/sites/default/files/resource/cma2023_L17_adv.pdf.

US Department of State (2025) *Minerals Security Partnership*. Washington, DC: US Department of State. Available at: http://www.state.gov/minerals-security-partnership.

USGS [United States Geological Service] (2025) *Mineral Commodity Surveys 2025*. Reston, VA: USGS. Available at: https://pubs.usgs.gov/publication/mcs2025.

Vakulchuk, R., Overland, I. and Scholten (2020) Renewable Energy and Geopolitics: A Review, *Renewable Energy and Sustainable Energy Reviews*, 122: 109547.

Van de Graff, T. (2023) Barrels, Booms and Busts: the Future of Petrostates in a Decarbonising World. In: Scholten, D. (ed) *Handbook on the Geopolitics of the Energy Transition*. Cheltenham: Edward Elgar.

Van de Graff, T. and Sovacool, B.K. (2020) *Global Energy Politics*. Cambridge: Polity Press.

Van de Loos, A., Langveld, R., Hekkert, M., Negro, S., and Truffer, B. (2022) Developing Local Industries and Global Value Chains: The Case of Offshore Wind. *Technological Forecasting & Social Change*, 174: 121248.

Van der Ploeg, F. and Rezai, A. (2020) Stranded Assets in the Transition to a Carbon-Free Economy. *Annual Review of Resource Economics*, 12: 281–98

Vantansever, A. (2021) *Oil in Putin's Russia: The Contests over Rents and Economic Policy*. Toronto: University of Toronto Press.

Vantansever, A. and Goldthau, A. (2025) The Political Economy of Breaking European Dependence on Russian Gas. *Resources Policy*, 109: 105696.

Veltmeyer, H. and Petras, J. (2014) *The New Extractivism*. London: Zed Books.

Vivoda, V. (2023) Friend-Shoring and Critical Minerals: Exploring the Role of the Minerals Security Partnership. *Energy Research & Social Science*, 100: 103085.

Vivoda, V., Matthew, R., and Andersen, J. (2025) Securing Defense Critical Minerals: Challenges and U.S. Strategic Response in an Evolving Geopolitical Landscape. *Comparative Strategy*, 44 (2): 281–315.

Wack, P. (1985) Scenarios: Unchartered Waters Ahead. *Harvard Business Review*, 63(6): 73–89.

Walter, D., Atkinson, W., Mohanty, S., Bond, K., Gulli, C., and Lovins, A. (2024) *The Battery Mineral Loop: The Path from Extinction to Circularity*. Basalt, CO: Rocky Mountain Institute. Available at: https://rmi.org/insight/the-battery-mineral-loop/.

Watts, M. (2004) Resource Curse? Governmentality, Oil and Power in the Niger Delta. Nigeria. *Geopolitics*, 9(1): 50–80.

Welsby, D., Price, J., Pye, S., and Ekins, P. (2021) Unextractable Fossil Fuels in a 1.5°C World. *Nature*, 597: 230–4.

Westphal, K., Pastukhova, M., and Pepe, J.M. (2022) *Geopolitics of Electricity: Grids. Space and (Political) Power*. Berlin: SWP, Research Paper 6. Available at: http://www.swp-berlin.org/10.18449/2022RP06/.

Wicks, M. (2009) *Energy Security: A National Challenge in a Changing World*: London: Department of Energy and Climate Change. UK Government.

Wilson, I. (2000) From Scenario Thinking to Strategic Action. *Technological Forecasting and Social Change*, 65(1): 23–9.

Wischer, G. and Bazilian, M. (2024) The Rise of Great Mineral Powers. *Journal of Indo-Pacific Affairs*, March-April: 162–84.

Wong, C.Y., Yeung, H.W.C., Shaopeng, H., Song, J., and Lee, K. (2024) Geopolitics and the Changing Landscape of Global Value Chains and Competition in the Global Semiconductor Industry: Rivalry and Catch-Up in Chip Manufacturing in East Asia. *Technological Forecasting & Social Change*, 209: 123749.

Woodroffe, N. (2024) *Responsible Change: How Governments Can Address Environmental. Social and Governance Challenges When Petroleum Assets Change Hands*. London: National Resource Governance Institute. Available at: https://resourcegovernance.org/publications/responsible-change-how-governments-can-address-environmental-social-and-governance.

World Bank (2025) *Development Indicators Database*. Washington, DC: World Bank. Available at: https://databank.worldbank.org/source/world-development-indicators.

World Bank Group (2004) *Striking a Better Balance – The World Bank Group and Extractive Industries: The Final Report of the Extractive Industries Review*. Washington, DC: The World Bank Group. Available at: https://documents1.worldbank.org/curated/en/961241468781797388/pdf/300010GLB.pdf.

REFERENCES

World Bank Group (2017) *The Growing Role of Minerals and Metals for a Low Carbon Future*. Washington, DC: World Bank Group. Available at: https://openknowledge.worldbank.org/entities/publication/4cdae3a6-3244-56e5-9de3-8faa5b6c88da.

World Bank Group (2020) *Minerals for Climate Action: The Mineral Intensity of the Clean Energy Transition*. Washington, DC: World Bank Group. Available at: https://documents.worldbank.org/en/publication/documents-reports/documentdetail/099052423172525564.

World Bank Group (2025) *Prosperity Data360*. Washington, DC: World Bank Group. Available at: https://prosperitydata360.worldbank.org/en/home.

WEC [World Energy Council] (2024) *World Energy Trilemma 2024: Evolving with Resilience and Justice*. London: WEC. Available at: http://www.worldenergy.org/transition-toolkit/world-energy-trilemma-framework.

WEF [World Economic Forum] (2019) *The Speed of the Energy Transition: Gradual or Rapid Change?* Geneva: WEF. Available at: http://www3.weforum.org/docs/WEF_the_speed_of_the_energy_transition.pdf.

WEF (2025) *Global Risks Report 2025*. Geneva: WEF. Available at: http://www.weforum.org/publications/global-risks-report-2025/.

Wrigley, E.A. (2010) *Energy and the English Industrial Revolution*. Cambridge: Cambridge University Press.

Yafimava, K., Ason, A. and Fulwood, M. (2025) *The EU Proposal to Ban Russian Gas Imports: Roadblock More Than Roadmap*. Oxford: OIES Paper NG: 199. Available at: https://www.oxfordenergy.org/publications/the-eu-proposal-to-ban-russian-gas-imports-roadblock-more-than-roadmap/.

Yeomans, M. (2004) *Oil: A Concise Guide to the Most Important Product on Earth*. New York: The New Press.

Yergin, D. (1990) *The Prize: The Epic Quest for Oil. Money & Power*. New York: Free Press.

Yergin, D. (2020) *The New Map: Energy. Climate and the Clash of Nations*. London: Allen Lane.

Yergin, D., Orszag, P., and Arya, A. (2025) The Troubled Energy Transition: How to Find a Pragmatic Path Forward. *Foreign Affairs*, March/April: 106–20.

York, R. and Bell, S.E. (2018) Energy Transitions and Additions? Why the Transition from Fossil Fuels Requires More Than the Growth of Renewable Energy. *Energy Research & Social Science*, 51: 40–3.

Zhang, Y., Jackson, C., and Krevor, S. (2025) The Feasibility of Reaching Gigatonne Scale CO_2 Storage by Mid-Century. *Nature Communications*, 15: 6913.

Zhou, J. and Månberger, A. (2024) *Critical Minerals and Great Power Competition: An Overview*. Stockholm: Stockholm International Peace Research Institute. Available at: http://www.sipri.org/publications/2024/research-reports/critical-minerals-and-great-power-competition-overview.

Index

References to figures appear in *italic* type;
those in **bold** type refer to tables.

A

adequacy **121**, 122–3
Africa 71, 76, 133
African Green Mineral Strategy 108
African Union (AU) 108
AI *see* artificial intelligence
alternative energy futures 30, **31**
Announced Pledges Scenario (APS) 28, 29, **29**, 73, 89, 95, 104
APS *see* Announced Pledges Scenario
Arab Spring (2010–11) 83
Archipelagos 23, 32
Arctic resources and Arctic-2 liquefied natural gas plant 65
artificial intelligence (AI) 22, 23, 114, 119, 120, 136, 137
Asia 45, 57, 66, 67, 83, 113, 125
Asian companies and demand 62, 63, 117, 131
Asian Financial Crisis 51
Asian super grid 125
AU *see* African Union
Australia 47, 102, 107
autocracy 83, 94
Azerbaijan 55, 61

B

Bahrain 84, 85
Baltic Sea and seabed 60, 63, 124
Baltic States and coast 60, 65, 125
batteries and electric vehicles 98, 114–19
BECCS *see* biomass and carbon capture and storage

Beijing summit meeting (September 2025) 67
Belarussian grid 60, 125
Belt and Road Initiative (BRI) 97, 125
Beyond Oil and Gas Alliance (BOGA) 132
biomass and carbon capture and storage (BECCS) 24
Bloomberg(NEF) **31**, 32, 33
BOGA *see* Beyond Oil and Gas Alliance
BP **31**, 32, 33
BRI *see* Belt and Road Initiative
BRICS bloc *see* Brazil, Russia, India, China and South Africa bloc
Bridges scenario 33
Brazil, Russia, India, China and South Africa (BRICS) bloc 107
Build Your Dreams (BYD) 116, 117

C

Canada 53, 71, 107, 126
carbon budget, global 89–94, 131
carbon capture and storage (CCS) 12, 33
technology 95
carbon dioxide (CO_2) 3–5
emissions 9, 12, 33, 38
carbon dioxide equivalent (CO_2e) 4
carbon efficiency 93
Carbon Tracker Initiative (CTI) 73, 89, 92, 93

Carter Doctrine 52
CATL *see* Contemporary Amperex Technology Co. Ltd.
CCS *see* carbon capture and storage
Central Asian gas 55
ChatGPT 4.0 32
Chile 102, 117
China 45, 53, 55, 56, 62, 65, 67, 96, 97, 101, 102, 103, 105, 107, 110, 111, 112, 113, 114, 115, 116, 117, 118, 119, 124, 125, 132, 135, 136, 137
Chip Wars 113
clean tech global production networks, de-risking of 135–6
clean tech supply chains 111–19, 134
climate action plans *see* nationally determined contributions
climate change 2–5, **6**, 13, 16, 101, 130
Climate Change Conferences *see* COP(s)
coal, geopolitics of 6–7, 45–6, 131, 132
Cold War 20, 52, 97
Comecon (Council for Mutual Economic Assistance) 59
conflict and democracy **48**, 50–1, 52–7, 77–8, 108–10, 133, 138
Contemporary Amperex Technology Co. Ltd. (CATL) 116, 117
COP(s) (Climate Change Conferences) 11, 25, 28, 80, 137
 -28 (Dubai, 2023) 1, 18, 92
 -29 (Azerbaijan, 2024) 1, 11
 -30 (Brazil, 2025) 24
Council for Mutual Economic Assistance, *see* Comecon
COVID-19 pandemic 27, 51, 62, 119
Crimea, Russia's occupation of 61, 63

critical materials/minerals
 challenge of 96–111, **99**, *102*
 Norwegian Institute for International Affairs / International Renewable Energy Agency composite list of for energy transition **100**, 100–1
 supply chains, diversifying of 134–5
Critical Materials programme 110
crude oil prices ($US), influence of geopolitical and economic events on *51*, 51–2
CTI *see* Carbon Tracker Initiative
Current Policies 28–9
Current Trajectory 32
cyberattacks 124

D

DAC *see* direct air capture
decarbonization 98, 118, 119, 132, 136
Delayed Transition 32
Democratic Republic of the Congo (DRC) 102, 103, 107, 110
direct air capture (DAC) 12, 24
Discord scenario 32
diversification 78, 80, 84–9, 102
domestic energy prices and consumption 78–9
DRC *see* Democratic Republic of the Congo
Dutch Title Transfer Facility (TTF) gas price *62*, 62–3

E

Economic Transition scenario 32
economies
 developed, high-energy, challenges to 9, 11, 12
 developing, challenges to 9, 11, 12, 49
 emerging, challenges to 9, 11, 12
 knowledge 87

INDEX

resource-abundant and non-resource-abundant 76
see also income levels
EI *see* Energy Intelligence
electric vehicle (EV)
  battery supply chain, simplified 114, *115*
  market/outlook 116, 118
  sales/uptake 32, 33
electric vehicles (EVs)
  batteries and 114–19
  government promotion of 98
  technology 95
  uptake of 132
electricity integration 121, 122–4, 126
electricity security 96, 119–27
  ensuring of 136–7
  terms and definitions, key 120, **121**, 122
  three 'i's of 121–7 *see also* electricity integration; grid interconnection; intermittency
electricity-based technologies 95
electrification 15, 40, 96, 119, 127, 128, 137
emissions gap and climate change financial risks 24–7, *25*
Emissions Gap Report (UNEP) 24
energy consumption 6, 29, *30*
energy demand reduction 15
energy-dependent states 73
Energy Dominance 36
energy efficiency 15
energy end-use 14, 15
energy futures 17, **31**, 34–5, 138
energy geopolitics 36–9
Energy Intelligence (EI) **31**, 32, 33
energy modelling 23–4, 27–8
Energy Outlook 2024 **31**
Energy Perspectives 2024 **31**
energy poverty 65
energy pragmatism 42
energy security 2, 26, 32, 33, 37–8, 39, 68, 116

Energy Security Scenarios 22 25 23, **31**
  – Archipelagos and Sky 2050 22
energy system transformation (EST) 5–13, 16, 17, 18, 19, 27, 36, 37, 38–9, 40, 43, 129, 130, 133
  critical geopolitical challenges of 130, **130**
  and the geopolitics of oil and gas *see* fossil fuel geopolitics
  and transition, distinction between 12–13
Energy Transformation Scenarios – Waves, Islands, and Sky 1.5 22
Energy Transition Macro Outlook 2025 **31**
Energy Transition Outlook 2024 **31**
Energy Trilemma 11
ENGOs *see* Environmental NGOs
Environmental NGOs (ENGOs) 93
environmental, social, and governance (ESG) outcomes 77, 109
Equinor **31**, 32, 33
Equitable Resource-Based Industrialisation 108
ESG outcomes *see* environmental, social, and governance outcomes
EST *see* energy system transformation
EU *see* European Union
Europe 57, 58, 59, 60, 61, 62, 63, 64, 65, 66, 67, 83, 110, 112, 117, 118, 119, 125, 126, 131, 135
  see also European Union
European allies 52
European buyers 66
European Commission 66–7
European context 124
European Critical Raw Materials Act (2024) 107
European gas crisis 47
European grids 123
European markets 55, 56, 125

European neighbours 126
European original equipment manufacturers 117
European Union (EU) 7, 24, 32, 60, 61, 63–4, *64*, 65, 66, 107, 110, 112, 117, 118, 119, 124, 125, 126 *see also* Europe
Europe's gas import infrastructure 58–62, *61*, 63–4
EV *see* electric vehicle
evidence of diversification 84–9
EVs *see* electric vehicles
export restrictions 106
external shocks 106
extraction costs 93

F

financial assistance for developing countries 25–6
fiscal breakeven price 85
Floating Storage Regasification Unit (FSRU) Independence 60
Floating Storage Regasification Units (FSRUs) 65
fossil fuel and renewable energy geopolitical issues 47, **48**
fossil fuel consumption 9
fossil fuel demand 79
destruction 65, 132
fossil fuel dependency 76, 80
fossil fuel exports and exporters 69, 92
fossil fuel geopolitics 45–68, 131
fossil fuel incumbents, managing the 94
fossil fuel rents 133
fossil fuel system 42, 43 *see also* 1 *under* Horizon
fossil fuels 4, 12–13, 18, 28, 38, 39, 76, 80, 83
and conflict **48**, 50, 109–10
phasing out of 44, 92, 131–3
unextractable 91, 92, 93
see also coal, geopolitics of; global *under* carbon budget; natural gas, geopolitics of; oil, geopolitics of

fossil gradualists 17
friendshoring 106, 107, 112, 118
FSRU Independence *see* Floating Storage Regasification Unit Independence
FSRUs *see* Floating Storage Regasification Units
futures, other available 30–5

G

G20 (Group of 20) 39
Gas Exporting Countries Forum (GECF) 46–7
Gazprom 59, 60, 62, 64
GCC *see* Gulf Cooperation Council
GDP *see* gross domestic product
GEC Model 2024 scenarios 28, **29**
GECF *see* Gas Exporting Countries Forum
GeGaLo (Geopolitical Gains and Losses) index 76
geographical concentration
of clean technology manufacturing capacity 112, *113*
of production and processing 101–5, *102*, *104*
of refined products 102, *104*
of selected critical materials 101–2, *102*
geography of unburnable fossil fuels 91–2
geopolitical challenges of energy system transformation, critical **130**
Geopolitical Gains and Losses index *see* GeGaLo index
geopolitical power **48**
geopolitical security and tensions 37, **48**
geopolitics
of energy system transformation 35–43, *41*, 129
and the high-carbon transition 130–4

of the low-carbon transition 39, 95–128, 134–7
of renewables 38
German Government 60, 61, 63, 65
GHG(s) *see* greenhouse gas(es)
Gigatonnes of carbon (GtC) 90
Global Carbon Project 90
global energy consumption 6–7, 7, 9
Global Energy Perspective 2024 **31**
Global Financial Crisis 51
global gas crisis, Europe's 62–8
global greenhouse gas emissions 4, *5*
global primary energy consumption 6–7, **8**, 9
global production networks (GPNs) 111, 114, 115, 117, 118, 136
global value chains (GVCs), creation of 111
global warming potentials (GWPs) 3–4
Global Witness 108, 110
GPNs *see* global production networks
green extractivism and agenda 97, 98, 108–11
greenhouse gas(es) (GHG(s)) 3–4
absolute volumes of 7
emissions 2, 9, 11, 12, 24, 86, 93, 101, 135, 137
objectives (2030–2035) 104
grid interconnection 124–7, 129, 137
gross domestic product (GDP) 70, 78, 84
Group of 20 *see* G20
GtC *see* Gigatonnes of carbon
Gulf 78
Rentier States and declining oil rents 80–9, *86*
Gulf Cooperation Council (GCC) 80, 81, 82, 83, 84, 87
fiscal breakeven oil price in 85, **87**
Monarchies of the 81, 82, 83
Gulf Monarchies 81, 82, 83
Gulf producers 93
Gulf Wars
(1990) 54–5
(1991, 2003) 52–3
Guyana 55, 71, 76
GVCs *see* global value chains
GWPs *see* global warming potentials

H

H1 *see* 1 *under* Horizon
H2 *see* 2 *under* Horizon
H3 *see* 3 *under* Horizon
high-carbon transition, just and equitable 80, 133–4
Horizon 22, 23, 33
1 (H1) 41, 42, 129
2 (H2) 41, 42, 129
3 (H3) 41, 42–3, 129, 130
Hungary 64, 67, 117

I

Iberian blackout (May 2025) 123
IEA *see* International Energy Agency
IEEFA's European LNG tracker *see* Institute for Energy Economics and Financial Analysis's European LNG tracker
IMF *see* International Monetary Fund
income levels 9, 11
India 72, 107
Indonesia 102, 103, 117
Industrial Revolution 5, 16
Inflation Reduction Act (IRA) (2022) 106
Institute for Energy Economics and Financial Analysis (IEEFA)'s European LNG tracker 65
Intergovernmental Panel on Climate Change (IPCC) 2, 3, 25, 29
intermittency 121–2
international competition **48**

International Energy Agency
(IEA) 27, 28, 29, 37, 39, 41,
73, 92, 95, 96, 102, 104, 111,
112, 116, 119, 120, 122–3,
127, 134–5
  energy scenarios of the 19,
  27–30, *29*, *30*, 89
  (World Energy) Outlook
  (2025) 28–9
international interdependence **48**
International Monetary Fund
(IMF) 11, 42, 76, 89
International Oil Companies
(IOCs) 92, 93, 94
International Renewable Energy
Agency (IRENA) 38, 100, 102,
105, 107
IOCs *see* International
Oil Companies
IPCC *see* Intergovernmental Panel
on Climate Change
IRA *see* Inflation Reduction Act
Iraq 76, 81
IRENA *see* International
Renewable Energy Agency

J

Japan 107, 113, 115, 117
joint ventures (JVs) 115, 117
JVs *see* joint ventures

K

Kuwait 84, 85
  Iraqi invasion of 54

L

liquefied natural gas (LNG) 46,
47, 48, 57–8, 60, 62, 63, 65, 66,
67, 68, 84
LNG *see* liquefied natural gas
location, importance of **48**
low-carbon energy system 42, 43,
45, 47, 68
low-carbon technologies, mapping
of minerals with 98, **99**

M

MAD *see* mutually
assured destruction
'Made in China 2025' industrial
strategy 112
McKinsey & Co **31**, 32
MENA countries *see*
Middle Eastern and North
African countries
Messy Mix 39, 42, 43, 44
  managing the 129–38
Messy Transition 35
methane (CH_4) 3, 4
Middle East 52, 71, 72, 76, 81, 92
Middle Eastern and North African
(MENA) countries 81, 133
Middle Eastern states 76, 81
Mineral Security Partnership
(MSP) 108
minerals and mineral cartels/
power 47, 97, 106 *see also* critical
materials/minerals
mining sector 101, 103
moderate-demand scenario 73, 76
MSP *see* Mineral
Security Partnership
mutually assured destruction
(MAD) 127

N

NASA *see* National Aeronautics
and Space Administration
National Aeronautics and Space
Administration (NASA) 2, 3
National Oceanic and
Atmospheric Administration
(NOAA) 4
National Oil Companies
(NOCs) 82, 92, 94
nationally determined
contributions (NDCs) 24, 25
NATO *see* North Atlantic
Treaty Organization
natural gas, geopolitics of 7, 28,
46–7, 48–9, 57–68, 77

INDEX

Nazi Germany and expansionism 36, 54
NDCs *see* nationally determined contributions
NEOM 87
net zero, achieving 1, 28, **31**, 33, 40, 42, 137
Net Zero (Scenario) 33, 39
Net-Zero Emissions by 2050 (NZE) (scenario) 28, 29, **29**, 95, 104
NEVs *see* new energy vehicles
new energy vehicles (NEVs) 115
New Energy Outlook 2024 **31**
The New Pragmatism **31**
NGOs *see* non-governmental organizations
Nigeria 76, 78, 79
NIMBYism *see* Not In My Backyard
Not In My Backyard (NIMBYism) 105
NOAA *see* National Oceanic and Atmospheric Administration
NOCs *see* National Oil Companies
non-governmental organizations (NGOs) 96
Nord Stream 1 pipeline 60–1, 63
Nord Stream 2 pipeline 62, 63, 67
North Africa 54, 81
North America 48, 57, 117, 119, 126, 131
North Atlantic Treaty Organization (NATO) 101, 124
Norway 65, 71, 126
Novatek 60, 65
NUPI/IRENA composite list of critical materials for the energy transition *see* Norwegian Institute for International Affairs / International Renewable Energy Agency composite list of for the energy transition *under* critical materials/minerals
NZE *see* Net-Zero Emissions by 2050

O

OECD *see* Organisation for Economic Co-operation and Development
OEMs *see* original equipment manufacturers
oil, geopolitics of 28, 46, 48–9, 50–7, 77, 80 *see also* fossil fuels
oil demand, falling 76
oil-dependent states 70–1
oil-exporting countries 77
oil-producing states, major 72, *73*
oil production 6–7, 71–2, *72*
oil rents 70, *71*, 89
  as a percentage of GDP 84, **85**
  countries where more than 10% of GDP 70, *71*
  and revenue 54, 55 *see also* Rentier State Theory and model; rentier states
Oil Security Paradigm 52
oil-wealthy states 71–2
OPEC *see* Organization of the Petroleum Exporting Countries
OPEC+ 46, 52
operational security **121**
Organisation for Economic Co-operation and Development (OECD) 111
  member states 27, 107, 132
Organization of the Petroleum Exporting Countries (OPEC) 46, 47, 51, 89
original equipment manufacturers (OEMs) 115, 116, 117, 118

P

Panel on Critical Energy Transition Minerals 110
Paris Agreement (2015) 1, 2, 13, 22, 24, 25, 28, 30, 32, 33, 34, 41, 43, 91, 129, 132
Persian Gulf 52, 53, 54, 69
petrocultures 109
petrol prices 53

Petromania 78
Petrostate Toolkit 80, 86
petrostates 53, 56, 70, 73, **75**, 76, 133
Philippines 102, 103
pipelines, international 58–62, *61*, 63–4
Poland and the Polish Government 59, 60, 63, 125
Power of Siberia
 pipeline 62
 2 pipeline 67
Powering Past Coal Alliance (PPCA) 132
PPCA *see* Powering Past Coal Alliance
Presidential Orders 106
price security 37, 53
price volatility 77, 79, 83
primary energy consumption 6–7, 9
producer economies 45, 49, 70–80, 94, 133, 134
production
 carbon intensity of 93
 restraint of 80
production states 81
productivity, failure of 78
profitability and sustainability 94
Putin, President Vladimir 49, 56, 63, 66, 67

Q

Qatar 47, 58, 81, 84–5

R

Rapid Transition 34
rare earths elements (REEs) 98, 100, 102, 103, 105, 110
REEs *see* rare earths elements
refined products, geographical concentration of 102, *104*
renewable energy 9, 38, 47, **48**, 122–3, 132, 134
 see also solar PV technology and sector; wind energy and technology

renewable technologies 42, 76, 132
rent addiction and dependence 80, 85, 94
rentier bargain 89
Rentier State Theory and model 80, 81–4, 85
rentier states 54, 81–2, 83
REPowerEU programme 64, 119
reshoring 106–7, 112, 118
resilience 76, **121**, 124–5, 137
resource curse 69–80, 82, 94
 characteristics, causes and consequences of the 76–9
 and the high-carbon transition 79–80
 policy prescription to address the 79
resource rents and revenues 79, 80
resource scarcity **48**
resource Supercycle 53
resource wealth 69, 76–7, 78
resources **48**, 101 *see also specific resources, e.g.* oil
risks 82, 103
 geopolitical 105, 106, 112
 physical 26, 27, 80
 transition 26–7, 73, 76, 80, 92–3, 94
Rupture scenario 33
Russia 55, 56, 57, 61, 64, 73, 92, 107, 124
 as Energy Superpower 49
 –Europe gas relations 58–62, 63–4, 65–8, 126
 oil supply of 52
Russian Government 59
Russian grids 125

S

S & P Global **31**, 32, 33
S-curve 16, 17
sanctions on oil and gas exports 49–50, 63, 66
Saudi Arabia 83–4, 85, 87
scarcity 16, 49

INDEX

<1.5 °C scenario 33
scenario planning and energy's futures 19–24, 80
SDG-7 see Sustainable Development Goal 7: Affordable and Clean Energy
security of supply (and demand) 37, **48**, 59–60
shale 48, 49
Shell 23, **31**, 32
Shell Scenarios 20, 22, 23
Slovakia 64, 67, 126
Slow Down scenario 32
Slow Evolution scenario 32
solar PV (photovoltaic) technology and sector 95, 112, 122
South Korea 113, 115, 117
Sovereign Wealth Funds (SWFs) 87
Soviet economy 55
Soviet natural resources 59, 97
Soviet states 92
Soviet Union 36, 54, 55, 58, 59, 60
speed of energy transitions 13–18, 21, 68
SPR see Strategic Petroleum Reserve
Staple Trap 77
Stated Policies Scenario (STEPS) 28, 29, **29**, 104, 119
Stellantis 117
STEPS see Stated Policies Scenario
stranded assets 90
and divestment 92–4
Strategic Petroleum Reserve (SPR) 27
superexporters 72–3
superproducers 72–3
supertraders 72
supply chains 96–7, 111–19, *115*, 134–5
Surge 22, 23
Sustainable Development Goal (SDG) 7: Affordable and Clean Energy 28
SWFs see Sovereign Wealth Funds

T

TANAP see Trans-Anatolian Natural Gas Pipeline
technological learning and innovation 16–17
technologies, low-carbon 98, 101
terawatt-hours (TWh) 7, 9, 119
Tesla 116
(US) 116
three horizons
approach to system change 40–1
and energy system transformation 41, 42, *43*, 129
Title Transfer Facility (TTF) gas price see Dutch Title Transfer Facility (TTF) gas price
TotalEnergies **31**, 32, 33
traders of crude oil and oil products, major 72, **74**
Trans-Anatolian Natural Gas Pipeline (TANAP) 61
transhipment 65–6
transit states 59–60
transition(s)
high-carbon 13, 81, 92, 129, **130**
low-carbon 13, 15, 17, 18, 24, 40, 66, 73, 110–11, 112, 126, 128, 129, **130**, 133, 134, 135, 136
messy 133
Trends 32
Trump, US President 1, 36, 49, 67, 106, 108
Trump Administration 66, 98, 112, 114
TTF gas price see Dutch Title Transfer Facility gas price
TurkStream pipeline 61, 64
TWh see terawatt-hours

U

UAE see United Arab Emirates
UK see United Kingdom

Ukraine 107, 125, 126
  Russia's invasion of and war in 37, 47, 52, 56, 63, 114, 123–4
  transit of Russian gas and other exports through 60, 62, 64, 66, 67
UN *see* United Nations
uncertainties, critical 21, 22, 27
UNEP *see* Environment Programme *under* United Nations
UNFCCC *see* Framework Convention on Climate Change *under* United Nations
United Arab Emirates (UAE) 84, 85
United Kingdom (UK) 65, 107, 112, 123, 126
United Nations (UN) 2
  Declaration on the Rights of Indigenous Peoples 109
  Environment Programme (UNEP) 24, 90–1, 110–11
  Framework Convention on Climate Change (UNFCCC) (1992) 11, 137
  Secretary-General 110
United States (US) 47, 48, 50, 52, 53, 59, 61, 65, 66, 71, 105, 107, 108, 111, 112, 113–14, 118, 119, 126, 136
  –China trade war 112
  CHIPS and Science Act (2022) 113
  Department of Defense 97
  Energy Dominance 49
  foreign policy 52–3
  Government 34
  Inflation Reduction Act (2022) 32
  liquefied natural gas production and trade 27, 57–8, 63, 64–5
  original equipment manufacturers 117
  –Soviet competition 52
US *see* United States

V

Venezuela 55, 76
Virtual Economy 56
Vision 2030 87
vulnerability 125
  of petrostates 73, **75**
vulnerable countries 72, 73, 134

W

Walls scenario 32
Warsaw Pact 59
WEC *see* World Energy Council
WEO *see* World Energy Outlook
West 52, 56, 97, 98, 101, 105–8, 118, 134, 135, 136
Western governments and alliance 109, 114, 118
Western oil companies 55
Western technology and original equipment manufacturers 55, 116
Wicks, Malcolm 37
wind energy and technology 95, 112
Wood MacKenzie **31**, 32, 33
World Bank income groups, key indicators by 9, **10**, 11
World Bank Sustainable Development Goal 7 12
World Economic Forum, White Paper 13
World Energy Council (WEC) 11
World Energy Outlook (WEO) (IEA) (2024) 28–9, *30*
world oil production and price by IEA's 2024 WEO scenarios **88**, 89

Y

Yamal–Europe pipeline 60, 63, 65

www.ingramcontent.com/pod-product-compliance
Lightning Source LLC
Chambersburg PA
CBHW051546020426
42333CB00016B/2129